Physiological Disorders of Fruit Crops

NIPA® GENX ELECTRONIC RESOURCES & SOLUTIONS P. LTD.
New Delhi-110 034

Physiological Disorders of Fruit Crops

By
Savreet Sandhu
&
Bikramjit Singh Gill

NIPA® GENX ELECTRONIC RESOURCES & SOLUTIONS P. LTD.
New Delhi-110 034

NIPA® GENX ELECTRONIC RESOURCES & SOLUTIONS P. LTD.

101,103, Vikas Surya Plaza, CU Block
L.S.C. Market, Pitam Pura, New Delhi-110 034
Ph : +91 11 27341616, 27341717, 27341718
E-mail: newindiapublishingagency@gmail.com
web: www.nipabooks.com

For customer assistance, please contact
Phone: + 91-11-27 34 17 17
Fax: + 91-11- 27 34 16 16
E-Mail: feedbacks@nipabooks.com

ISBN: 978-81-19254-54-5

Composed and Designed by NIPA®.

CONTENTS

Chapter-1

Introduction

Physiological or abiotic disorders are mainly caused by changing environmental conditions such as temperature, moisture, unbalanced soil nutrients, inadequate or excess of certain soil minerals, extremes of soil pH and poor drainage. The distinction between physiological or abiotic disorders from other disorders is that they are not caused by living organisms (viruses, bacteria, fungi, insects, etc.), but they are the result of abiotic situations (inanimate) i.e. their agents are non-living in nature which causes deviation from normal growth. They results in physical or chemical changes in a plant which is far away from what is normal and is generally caused by an external factor. Non-infectious disorders in some cases are easy to identify, but others are difficult or even impossible to recognize. Most of them are non reversible once they have occurred. For the identification of physiological disorders it is important that one must know that:

a) Physiological disorders are often caused by the deficiency or excess of something that supports life or by the presence of something that interferes with life.

b) Physiological disorders can affect plants in all stages of their development.

c) They are non-transmissible because they occur without or in absence of infectious agents.

d) Plant reacts differently to the same agent and sometimes response is seen as a little reaction to death.

e) Dealing with physiological disorders often means dealing with the consequences from a past event.

f) Generally damaged and undamaged tissue is clearly demarcated.

g) Physiological disorders not only causes damage themselves but also serve as the 'open door' (entry) for pathogens.

Factors implicated in occurrence of physiological disorders include:

1. *Irradiance (intensity, photoperiod and spectral quality):* The availability of light either totally or partially, its quality and source is very important. Too much red or blue spectrum light can affect plants and their growth, depending on the situation and plant. The duration of light is also very important in the production of plants where it must have specific day/night periods.

2. *Relative humidity (RH):* Low RH accompanied by high or excess of temperature will generally occur as a problem. Higher RH over longer periods of time like in a greenhouse or wet climate areas can also support establishment of certain disease organisms.

3. *Atmospheric conditions:* Atmospheric conditions overall play havoc with fruit plants in some parts of the world. One of the most notable is acid rain/snow that with time causes the water and soil to become too acidic for growing any plant. Too much cloud or heavy smog can also show a negative affect on plants causing low light intensity situations.

4. *Carbon dioxide concentration:* Too high or low levels of CO_2 can also affect plant growth and this situation is commonly confronted in closed atmosphere places like greenhouses and it must be controlled for proper plant growth.

5. *Heat stress / scorch injury:* This is one of the major physiological disorders especially in hot climates and it is caused by injury from direct sun exposure/intensity or just simply from too hot climate.

6. *Winter/cold injury and frost:* Depending on severity, frost injury may cause browning of fruit tissue, deformation, puckering and

damaging of fruit parts to complete death of a fruit. The frost damaged parts will generally show evidence very quickly. Winter injury can also occur in the form of ice and/or snow damage. Injury in the form of slight damage to severe injury or death can occur from excessively low soil temperatures also.

7. *Wind injury:* Wind injury can also aggravate cold injury or winter injury, especially if the humidity is low at the time of the wind/ cold period. This is especially hard on buds or tender fruits. Wind damage can be as simple as plant parts rubbing together causing surface scarring which also serve as entry points for pathogens. Wind will create more evaporation from the leaf and fruit surface on a hot day, possibly aggravating a heat stress factor.
8. *Chemical injury:* Any kind of foreign chemical applied in the wrong dosage or at the wrong time is capable of doing physical damage to the fruit plants. Most chemical injuries will come from pesticides applied at too high rates, at wrong time or during very hot parts of the day. Damage from chemical injury may appear as red, yellow or brown spots on the fruit skin, leaf tips turning brown, stunted or misshapen fruits, to overall browning and death of a plant.
9. *Mechanical or physical injury:* This occurs where plants are physically damaged by people, wind, animals, equipment, etc. This is important as every site of tissue injury is a potential entry point for disease.
10. *Physical soil problems:* The selection of a good site for growing is very important as soil can have a physiological effect on the plant, which can take many forms. Compacted soil allows water to percolate slowly into the soil causing saturated conditions resulting in soil deformation. Compaction can be due to low organic matter (OM) in many soils which makes the work difficult on such soils.
11. *Water stress:* This is the situation when the plant gets to much or not enough water to function properly. The water stress situations include heat where the plant cannot take up water fast enough for its needs, improper maintenance of irrigation equipment such as plugging of sprinkler heads for prolonged period of time or breakdown with long repair times. Further, it includes competition from other plants such as large trees which have massive root systems often reaching far in each direction. Such

trees are exhaustive and have little water in the soil, making it hard or impossible for other plants to retrieve it. Water stress is also caused by too much rain and can be enhanced on sloped sites where water from rain or irrigation runs to a low spot causing super saturated areas or standing water.

12. *High salts (Electrical conductivity):* High concentration of salts is another problematic area that can cause symptoms which appear as physical problems, but they are actually caused by chemical imbalance in the soil. Small, stunted, slow growing plants or leaf tip burn could be typical symptoms related to this malady. Occurrence of high salts in soil might be due to excessive fertilizer usage or naturally because of soil chemical conditions.
13. *Improper planting techniques:* There are a number of things that fall into this category such as too deep planting either as seeds or plants causing poor or no growth, planting too late in the year, girdling of roots, lack of inspections or improper root pruning at planting time, etc.
14. *Nutrient deficiency or excess (Imbalances):* Imbalance of different nutrients will give different reactions and an excess of one nutrient can make another one unavailable or non-functional. Every plant reacts different to excessively high or low levels of nutrients, some reactions being quite evident, others less. Plants can be stunted, deformed or symptoms may occur as leaf tip or leaf margin burn.
15. *Genetic/hereditary factors:* Although this is an internal factor and not an external cause, plants can do weird things when its own genetic system goes awry. This can cause many factors from distorted stems (fasciation), colour changes (discolouration), multiple stems etc.

Development of disorders during post-harvest ripening and storage of fruits depends on a range of factors. Pre-harvest environmental conditions and orchard practices modify incidence of disorders induced specifically by storage conditions such as low temperature or high CO_2. In terms of general postharvest product quality, this preharvest impact is implicit in the use of maturity indices for optimizing harvest times. Some preharvest factors may be effective through influencing fruit maturity. However, many direct effects of growing conditions and fruit development on post-harvest disorder incidence may be quite independent.

To understand the relationships between fruit development and post-harvest quality it is useful to divide disorders into those which are predetermined on the tree and those which are specifically induced by storage conditions yet modified by preharvest factors. The division may be somewhat artificial, although still useful, in that some disorders fall in both categories. Pre-harvest factors which may predispose fruit for subsequent disorder development are dominated by position of the fruit on the tree, characteristics of the fruiting site, crop load, mineral and carbohydrate nutrition of the developing fruit, water relations and response to temperatures. Expression of disorders which are the direct result of storage conditions, such as low temperature and high CO_2, will still be modified by similar factors. An understanding of these factors allows modification of fruit development to optimize storage quality and development of methods for predicting disorder risk.

Fruit picked either prematurely or too late, are more susceptible to post-harvest physiological disorders than are picked at the proper stage of maturity. "Physiological disorders" sometimes termed as abiotic problems, reflects the fact that the injury or symptom seen on the plant such as reduced growth or crown dieback, is ultimately due to the cumulative effects of the causal factors on the physiological processes needed for plant growth and development. When a tree is affected by severe drought, for example, water stress, it will result in closure of the pores or stomata on the leaf. This conserves water in the leaf but also reduces the rate of photosynthesis and the ability of the plant to produce sugars for growth. If the drought stress occurs during hot weather, stomata closure also limits the cooling effect of transpiration, and leaf scorch may occur. Nutritional imbalances also limit growth by reducing photosynthetic rate and other physiological and metabolic processes. The cumulative and subtle nature of many physiological disorders can often make them difficult to diagnose. More often, however, diagnosing abiotic problems requires careful consideration of plant and site factors through a process of elimination to determine the source and potential remedy for the problem.

Both quantitative and qualitative losses occur in fruit crops before harvest and between harvest and consumption. Qualitative losses such as loss in edibility, nutritional composition, calorie value and consumer acceptability of the products are much more difficult to assess than the quantitative losses. Pre-harvest and post-harvest physiological losses vary greatly among commodities, production areas and seasons. Postharvest losses are more profound than pre-harvest losses. Estimates of post-harvest losses in developing countries like India vary greatly from 1 to 50 per cent or even higher. Growers can produce large quantities of good quality fruits but if they do not have a dependable, fast and equitable means of getting such commodities to the consumer, losses will be extensive. There is desperate need of improvement in pre and post-harvest facilities and sanitation to decrease such losses.

As many post-harvest losses are direct results of pre-harvest factors so fruits obtained from inappropriately irrigated and fertilized trees or fruits which are generally of poor quality before harvesting, can never be improved by post-harvest treatments. Very often the rate of commodity loss is faster if the quality at harvest is below standard. Thus, the processes in the attainment and maintenance of quality from production, harvesting, handling and marketing must be considered a unified system. Therefore, the success of preserving the fresh quality produce demands the control of each step in the system which is a chain of interdependent activities.

❑❑❑

Chapter-2

Aonla

Scientific Name : *Embilica offincinalis* Gaertn.

Family : Euphorbiaceae

Aonla, indigenous to Indian subcontinent popularly known as Indian gooseberry (Amla), is a fruit with immense nutritional and medicinal value. In fact, aonla, in its processed form is very popular among the social elites. Its fruit is valued very high amongst the indigenous medicinal fruits and has even been recommended by the ancient Ayurvedic system for sound health. Aonla is the second largest source of Vitamin C (600 mg/100 gm) among all the fruits, after Barbados cherry and this is 10 times the amount of Vitamin C present in orange. Vitamin C plays a major role in the health of the teeth and gums. In addition to this, it increases the resistance against pathogenic attacks. Aonla is also rich in

Fruit necrosis

fibers (3.4%) that are important for the smooth flow of ingested food in our alimentary canal. It also contains carbohydrates (14%) and vitamin B_1 (30 mg/100g of fruit weight). Minerals like iron, calcium and magnesium, which are crucial for the various metabolic reactions in a human body, are also present in ample quantities in aonla. Most of the fruit is used for preparing 'Murabba', chutney and pickle. It is also used in the preparation of Chavanprash.

Owing to its hardy nature, suitability to various wastelands, high productivity and nutritive ingredients besides having therapeutic values, aonla has become an important fruit. Aonla is not merely a source of nutrients and medicine, its cultivation is also highly remunerative for small and marginal farmers. The cultivation of aonla is confronted with physiological disorders like fruit necrosis and fruit drop which are discussed in detail below.

1. Fruit necrosis

Fruit necrosis is especially a problem in cultivar 'Francis' where necrosis is noted upto 60-80 per cent and it is followed by cultivar Banarasi.

Symptoms

The symptoms start with the browning of the innermost part of mesocarpic tissue at the time of endocarp hardening which later extends towards the epicarp resulting into brownish black appearance of the flesh in the second and third week of October. Depending on the severity of the disease, mesocarp of the affected fruits turns black from brown forming dark spots. The dark spots become corky and gummy in the later stages.

Causes

Internal fruit necrosis is associated with boron deficiency during the fruit development.

Control

Infection has not been noticed on other aonla cultivars like Chakaiya, NA-6 (a seedling selection from Chakaiya) and NA-7 (a seedling selection from Francis). Hence, the cultivation of these varieties needs to be encouraged. It could be controlled with three sprays of borax at 0.6 per cent at fortnightly interval commencing from early September to October at 10-15 days interval.

2. Fruit drop

It is a serious problem in aonla which influences the final yield of crop. Three waves of flower and fruit drop are observed in aonla which are as follow :

a. *First wave:* More than 70 per cent flowers drop within three weeks of flowering which is normally due to lack of pollination.

b. *Second wave:* The second drop is that of young fruitlets at the time of dormancy break.

c. *Third wave:* The third drop of fruits is due to embryological and physiological factors and it is spread over the entire period of the fruit development. It starts from later half of August and continues upto harvest.

Causes

Fruit drop results due to many factors such as:

i. Dry spell

ii. Imbalance of hormones

iii. Improper nutrition

iv. Fluctuation in temperature

v. Variety and age of tree

vi. Number of developing fruits

Also, delay in harvesting results in heavy fruit drop particularly in Banarasi and Francis cultivars.

Control

To secure heavy fruit set and reduce fruit drop, watering at bi-weekly intervals of mature bearing plants is necessary from April to June. Also, growth regulator sprays are found useful in controlling fruit drop.

❑❑❑

Chapter-3

Apple

Scientific Name : *Malus domestica*

Family : Rosaceae

Rusetting

Delicious and crunchy apple fruit is one of the most popular fruit favoured by health conscious, fitness freaks who believe in "health is wealth". It is one of the popular wonderful fruit that contains a long list of essential nutrients that in the true sense are indispensable for human health. The antioxidants in apple have many health promoting and disease prevention properties; thus justifying the adage *"An apple a day keeps the doctor away"*. Apples help in controlling heart diseases, weight loss and cholesterol levels. Apples do not have any cholesterol instead they have fibre which reduces cholesterol by preventing reabsorption and thus are low in calories. Unpeeled apples provide plenty of nutrients which are present just under the skin. Apples are a good source of minerals like potassium, calcium, magnesium and iron, vitamins of B-complex such as riboflavin and pyridoxine (vitamin B_6), vitamin C, folic acid and beta carotene. Apart from this they also contain appreciable quantities of tartaric acid that gives tart flavour to them.

But such a nutritive fruit is also confronted by many physiological disorders that occurs due to factors like environment, nutrient deficiency, storage conditions, etc. which reduces the market value of the fruit. In apples, such disorders include scald, bitter pit, cork spot, Jonathan spot, water core, internal breakdown, fruit drop, alternate bearing etc. They can appear during the growing season or after harvest when the fruits are being stored and they affect the appearance and usability of the fruit. In some instances, pre-harvest stress conditions predispose the fruit to develop symptoms of a disorder only after picking or storage. Thus, it is important for both growers and packers to be aware of these potential problems.

1. Scald

This storage disorder is of serious concern for apple growers. Susceptibility to this storage disorder varies with the variety of apple, environment and cultural practices. Granny Smith, Rome Beauty, Delicious, Winesap and Yellow Newton varieties are very susceptible whereas Gala and Fuji are moderately susceptible. The immature fruits are most susceptible to scald which is aggravated by warmer temperatures in storage. Light scald fruits may be severely affected during marketing. Storage scald has been a serious concern for apple growers as long as apples have been stored and marketed commercially.

Symptoms

Light mottling on greener surface of fruits are initial symptoms of scald. Irregular brown patches of dead skin develop within 3 to 7 days due to warming of the fruit after removal from the cold storage. Darkening becomes more severe with elapsed time and ultimately extends to red surface also. Scald usually affects the skin only but in severe cases it may extend to fruit flesh. Symptoms may be visible in cold storage when injury is severe. The warm temperatures do not cause the scald but allow symptoms to develop from previous injury, which occurred during cold storage.

Causes

The browning of the epidermal and hypodermal cells of the fruit seems to be associated with an oxidation product coming from

α - farnesene formed in the waxy coating of the fruit. Incidence and severity of scald is favoured by following factors.

i. Hot and dry weather before harvest.

ii. Immature fruit at harvest.

iii. High nitrogen and low calcium concentrations in the fruit.

iv. Inadequate ventilation in storage rooms or in packaging boxes.

Control

a. The most common method used to control scald is application of an antioxidant immediately after harvest.

b. Commonly used antioxidant is Diphenylamine (DPA). Ethoxyquin is also effective for some varieties, but can cause damage to other apple varieties.

c. Antioxidants should be applied within one week of harvest for maximum control. Spray of $CaCl_2$ (2-3%) two weeks before harvest is very effective for controlling scald effectively.

d. Harvesting at proper maturity and ventilation in cold storage helps in reducing the scald incidence

2. Bitter pit

Bitter pit is considered at the most serious disorder of apple which reduces the fresh market quality of fruit. This physiological disorder is influenced by climate and orchard cultural practices. Young trees that are just coming into bearing are the most susceptible. The fruits of Northern Spy, Golden Delicious, Yellow Newton and Gravenstein are most susceptible although it affects almost all apple varieties.

Symptoms

Small brown lesions of 2-10 mm in diameter (depending on the cultivar) develop in the flesh of the fruit. Bitter pit is characterized by small sunken spots on the fruit surface which are more prevalent near the blossom-end. The tissue below the skin becomes dark and corky. At the initial stage small water soaked areas appears which after loss of water shrink and turn brown and ultimately become brown and corky due to the dead tissue. Unlike the name, these corky tissues are never bitter in taste. The spots generally turn darker, become more sunken than the surrounding skin and get fully developed after one or two months in storage.

Causes

There are many factors which have been found associated with bitter pit. These are:

i. Nutrient imbalance particularly low level of calcium which impairs the selective permeability of cell membranes leading to cell injury and necrosis.

ii. Heavy doses of nitrogenous fertilizers as it results in lowering the soil pH or inducing excessive vigour.

iii. Excessive shedding and heavy pruning as severe dormant prunong would result in a light crop and large fruit.

iv. Early and over thinning of fruits increases bitter pit

v. Irregular water supply.

vi. Harvesting of immature fruit, as early picked fruit tends to develop more bitter pit symptoms.

Control

Bitter pit can be controlled by:

a. Avoiding excessive doses of nitrogenous fertilizers.

b. Thinning of fruits should be done judiciously.

c. Annual bearing

d. Maintaining moderate tree vigour and smaller fruit size.

e. Harvesting mature fruit

f. Calcium sprays prior to harvest and calcium dips before storage control the incidence of bitter pit. The plants should be sprayed 45 days prior to harvest followed by repetition of spray after 15 days. The post-harvest dip for 1-2 minutes should be given before storage. Storage of fruits at 32-35°F under high humidity (85-95%) also controls pitting.

3. Internal browning (Brown heart)

The disorder is associated with later/harvested, large and over mature fruit and with high CO_2 concentrations in storage. Because internal browning is not detectable externally, except in very severe cases, affected fruit can be discovered by buyers or consumers, thereby, affecting future confidence in the product. Although, fruit susceptibility varies from season to season and orchard to orchard, considerable losses of controlled atmosphere stored fruit can result from this disorder. The reasons for this variability are unknown but it is suggested to be associated with cool, wet weather and high nitrogen fertilization. Internal browning is very prominent in Cox's Orange Pippin and Jonathan apples.

Symptoms

Internal browning is characterized by brownish streaks radiating into flesh from the core. These brown areas have well defined margins and may include dry cavities resulting from desiccation. Symptoms can range from a small spot of brown flesh to nearly the entire flesh being affected in severe cases. When the entire apple is affected, a margin of healthy, white flesh usually remains just below the skin. Symptoms develop early in storage and may increase in severity with extended storage time.

Causes

Internal browning occurs due to following reasons.

i. *CO_2 injury to the apple:* Injury incidence and severity increase with increasing concentration of CO_2 in the storage atmosphere. Variability in susceptibility of apple varieties and in apples of different maturity levels may be due to anatomical differences (cell size, size of intercellular spaces) rather than biochemical differences.

ii. *Large and over mature fruits:* Later harvested fruits generally have greater susceptibility to internal browning

Control

Following practices can aid in controlling internal browning in apple:

a. Avoid harvesting over mature fruit.

b. Harvest at the optimum maturity, especially for controlled atmosphere storage.

c. Maintain CO_2 concentrations below 1% in controlled atmosphere storage and air storage.

d. Assure good air circulation in storage rooms to prevent formation of higher CO_2 concentration pockets. Proper temperature management and good ventilation will prevent buildup of CO_2.

e. Appearance of disorder is less at 1°C in cold storage, so control the temperature.

f. Avoid heavy wax coatings and thoroughly and rapidly cool fruit after waxing and packaging.

4. Cork spot

Cork spot is a common problem of fruit deformity in apples. Among the most affected varieties are Red and Yellow Delicious, Stayman Winesap and York Imperial.

Symptoms

The initial symptoms of this physiological disorder appear as small blushed area on the skin of the fruit above the affected brown spot in the flesh of the fruit. On the outside, affected apples often appear knobby, pitted and deformed. The affected tissue is usually much harder than the healthy tissue.

Causes

Low calcium in the fruit is one of the basic cause of cork spot. It is essentially nutritional problem; however, number of factors can contribute to this disorder as discussed below:

i. *Poor nutrition:* This may be caused by higherlevels of nitrogen and lower levels of calcium and boron. These deficiencies are found responsible for development of cork spot.

ii. *Unfavorable soil conditions:* These can result from compaction, excessive acidity (low pH), and especially lack of moisture in the soil.

iii. *Excessive tree vigour:* This is often the result of heavy pruning, which stimulates a lot of vegetative growth at the expense of fruit. The calcium needed by the fruit gets diverted to the vigorous new growth.

iv. *Low fruit production:* This situation can also lead to cork spot. Poor pollination and late frost damage can cause a lower yield. The fruits tend to outgrow their supply of calcium when there are few apples on the tree.

Control

a. Proper nutrient management especially boron and calcium helps in preventing this disorder.
b. Keep the soil properly limed to maintain the pH around 6.5.
c. Avoid heavy doses of nitrogen application and apply this under the spread of the branches.
d. To supply the immediate needs of an apple crop, spray calcium chloride (0.5%) directly on the fruit and foliage during the growing season.
e. Injecting calcite lime in the trees could reduce its incidence.

5. Water core

This physiological disorder is difficult to detect externally unless symptoms are very severe. Incidence varies from season to season but can be very severe if temperature remains high. Large size fruits are mostly susceptible to this type of disorder. Susceptible varieties include Jonathan, Delicious, Stayman, Winesap, Granny Smith and Fuji. Water core is found more on the exposed side of the apple and may be associated with sunscald. The water soaked appearance of water core affected fruit results from the accumulation of sorbitol-rich solutions in the intercellular spaces.

Symptoms

The symptoms of water core develop rapidly when the fruits are kept on the trees for longer time. Disorder at pre-harvest stage results in the development of water soaked regions in the flesh which are hard, glassy in appearance and only visible externally when infection is very severe. The fruits may smell and have a fermented taste when severely affected. Water soaked areas are found near the core or on the entire apple. If symptoms are mild to moderate, they may disappear completely in storage. However, when water core is severe, internal breakdown can develop.

Causes

Water core is promoted by following factors:

i. Large fruit and high leaf to fruit ratio.
ii. High fruit nitrogen and boron.
iii. Low fruit calcium.
iv. Excessive thinning.
v. High light exposure.
vi. Ethrel sprays.
vii. Girdling of trunks and limbs.

Control

a. The most effective way to reduce the incidence is to avoid delayed harvests.
b. As fruits approach maturity stage, samples of fruit should be examined for water core development. Fruit should be harvested before water core develops extensively.
c. Fruit lots with moderate to severe water core symptoms should not be placed in controlled atmosphere (CA) storage but should be marketed quickly.

6. Sunburn

This physiological disorder occurs due to intense heat of the sun and fruits on the southwest side of the trees are generally affected with this malady. The greatest problem occurs when fruits are suddenly exposed to high temperatures and intense sunlight when the weather rapidly shifts from being cloudy and cool for many days to sunny and warm. Granny Smith and other light skinned apple varieties are more sensitive to sunburn. Two types of sunburn have been distinguished in apple, the first type is "sunburn necrosis" and the second type is "sunburn browning". It can drastically reduce yield and fruit quality.

Symptoms

Initial symptoms are white, tan or yellow patches on the fruits exposed to the sun. With severe skin damage, injured areas of the fruit can turn dark brown before harvest. These areas may become spongy and sunken. Fruit exposed to the sun after harvest can develop severe sunburn. The damage occurs mainly in the surface and sub-surface layers.

Causes

The greatest problem occurs mainly due to following factors.

i. Sunburn occurs when air temperature and the number of sunny hours are high during the ripening period.

ii. Sunburn also occurs when cool or mild weather is abruptly followed by hot, sunny weather.

iii. Sudden exposure of fruit to high temperatures and intense sunlight promotes sunburn.

iv. A gradual increase in temperature and solar radiation is another reason.

v. Heavy crops that cause branches to bend over mid-season can increase sunburn incidence as a sudden exposure to heat and sun promotes sunburn development.

vi. Water stress can increase the incidence of sunburn.

Control

To avoid the incidence of sunburn following control measures are helpful:

a. The best method is to avoid sudden exposure of fruit to intense heat and solar radiation.

b. Proper tree training and pruning practices should be followed.

c. Summer pruning must be carefully done to avoid excessive sunburn.

d. Pruned orchards should be regularly irrigated to reduce heat stress.

e. Overhead sprinkling and white washing of trees reduces sunburn.

f. Careful sorting to remove affected fruit upon packing is the only solution once the injury has occurred.

7. Russeting

Russeting of apples in humid environments is of major concern to the fruit growers. It occurs shortly after petal fall. The apple cultivars, which have thin cuticle, are more susceptible to russeting. It is commonly noticed on exposed fruits than on fruits remaining in shade. Russeting leads to rupturing of the fruit skin and development of cracks. The fruit of younger or vigorously growing trees seem more prone to russeting than older and slower-growing trees.

Symptoms

Russeting is a brown, corky netlike condition on the skin of apples. It may appear only on small portion of each fruit or it may cover its whole surface. Severe russeting may be accompanied by fruit cracking which usually renders the fruit useless. When russeting occurs early, the skin becomes thin and tight. It can't stretch with the developing, enlarging fruit, so it cracks. These cracks into the flesh allow fungi and bacteria to enter and rot the fruit. When russeting occurs later during the fruit development, cracks don't form but the skin becomes brown, dull, and rough.

Causes

Russeting has been associated with large number of factors which are as follow.

i. Frost during floworing or at the early fruit formation stage promotes russeting.

ii. Specific environmental conditions such as high temperatures and humidity also favours russeting.

iii. Excessive nitrogen application to the trees is another cause.

iv. Damage from harsh chemicals and caustic sprays is one of the factor. Spray-induced russeting often occurs only on the exposed side of apples. The use of pesticide formulations called emulsifiable concentrates (ECs) is more likely to result in russeting than using wettable powders (WPs).

v. Infection by certain fungi, bacteria and viral organisms also increase the incidence of russeting.

Control

Following practices appear to be beneficial in controlling russeting of apples:

a. Selection of less susceptible cultivars.

b. Adequate irrigation, manuring and effective pest management can reduce russeting.
c. Avoid spraying emulsifiable concentrates (ECs).
d. Do not apply chemicals during slow drying conditions, high humidity, or temperatures above 90°F.
e. Check for uneven spray distribution.
f. Also control powdery mildew.
g. Prune properly to encourage good air circulation and speed fast drying after rains.
h. Avoid heavy crop load so encourage fruit thinning.

8. Jonathan spot

The disorder was first described on Jonathan apple and is perhaps most common on this variety. Other varieties, however, such as Yellow Newton, Rome Beauty, Gravenstein, Northern Spy and Golden Delicious may be affected. Its most serious characteristic is a tendency to develop to such an extent in transit or storage that marked damage occurs in fruits that were apparently in good condition when shipped or stored.

Symptoms

Jonathan spot is a disorder of Jonathan apples characterized by irregular small brown to black spots on the skin which originates at the lenticels. Two distinct types of spots may occur on the Jonathan variety. One is a sharply sunken area at the lenticels that is very dark brown or black and about 1/8 of an inch in diameter. These spots penetrate the flesh somewhat below the surface of the skin. Other is superficial which is light brown, irregularly shaped that surround the affected lenticels and may cover an area up to 1/4 inch in diameter. Jonathan spot tends to occur on the blush side and occurs more at the stem half or around the middle of the apple than at the calyx end.

Causes

The causes of this physiological disorder are still not fully understood, but possible predisposing factors which have been suggested are as under.

i. Jonathan spot is common after a dry season.
ii. It is usually worse on large apples than on small ones.

iii. It is worse on late-harvested fruits than on those harvested at prime maturity.

iv. It occurs due to insufficient use of nitrogen fertilizer.

v. Delayed cooling and prolonged storage increases its incidence.

Control

a. Harvesting fruits at optimum maturity.

b. Prompt cooling, and sharp control of storage temperatures reduces aging and delays the onset of Jonathan spot.

c. Storage in controlled atmospheres of 2.5% to 5.0% carbon dioxide and 3.0% oxygen controls the disorder on Jonathan apples.

d. Spraying of calcium chloride (0.5%) before harvest decreases the incidence of this disorder.

9. Fruit drop

Fruit drop is a complex phenomenon which occurs due to large number of factors like variety, tree aspect, environmental conditions, nutritional condition of the plant, number of developing fruits, number of seeds in the developing fruits and hormonal imbalance particularly auxin. In apple it is a very serious problem and the magnitude of the problem can be realized by the fact that in apple hardly 1 per cent flowers take their fruits to maturity and others drop at different stages.

Most of the commercial varieties experience 3 phases of fruit drop, out of which first two drops viz., early drop and June drop do not cause substantial economical losses but the third drop viz., preharvest drop causes serious economic loss as it is related with the full grown marketable fruits. The three waves of drop are:

i. *Early drop:* It is considered natural and occurs due to lack of pollination, fruit competition, wind, tree aspect etc. It occurs shortly after petal-fall and may continue for 2 to 3 weeks. The fruit that falls during this period is pea-sized and may be the result of poor pollination. Lack of pollination may also be due to poor weather as most fruit trees are pollinated by bees and they are most active on sunny, warm days and there activity during cool, rainy weather living low which results in poor pollination during the bloom period leading to fruit drop.

ii. *June drop:* It occurs during the summer months when there is scarcity of moisture. The fallen apples are approximately 1/2 to

1 inch in diameter. The second shedding of fruit is often due to the competition for food, water and nutrients among the developing fruits. This drop is mainly caused by moisture stress and environmental conditions. Hot, dry weather in late spring contributes to fruit drop.

iii. *Pre-harvest fruit drop:* It causes serious economic losses, as the mature marketable fruits abscise before harvesting due to reduction in levels of auxins or increase in ethylene levels in fruit. This drop is serious in early ripening varieties like Tydeman's Early and Red Gold ranging from 40-60% of drop, whereas mid season cultivars like Delicious group and Golden Delicious experience 15-20% of crop loss.

Control

a. The soil condition and disturbed water relations can be checked by maintaining soil moisture through irrigation and mulching.

b. The pre harvest fruit drop can be checked conveniently with the application of 10 ppm NAA before the expected fruit drop or 20-25 days before harvest. Application of 2,4-D has also been found to reduce pre-harvest fruit drop effectively.

10. Alternate bearing/ Biennial bearing

Biennial bearing is a common disorder of apple trees where they crop heavily in one year and then produce little or not found in the next year. Without a crop to support in any one year, trees use their resources to produce flower buds leading to tremendous blossom in the following year. The resulting heavy crop reduces the tree's resources so that little blossom is made in the following year. The apple cultivars Royal Delicious, Red Delicious and Golden Delicious have the tendency to bear heavy crop in a year followed by poor cropping season, setting a rhythm of irregular bearing. Once the pattern of irregular bearing sets in, the fluctuating rhythm of yield continues year after year. Biennial bearing plants of apple produce higher yields of small, poor quality fruit in the "on" year and low yields of large fruit prone to physiological disorders in the "off" year. Flowering promoters such as Ethephon may help boost return bloom after an "on" year, and floral inhibitors such as gibberellic acid (GA_3) may reduce bloom in the season following an "off" year, improving cropping consistency and orchard profitability.

Causes

Not only a single factor is responsible for this complex problem. It is due to consortium of many factors.

i. *Environmental and climatic factors:* Biennial bearing can also be started by frost destroying the blossom in spring.

ii. *Insufficient moisture:* It is most likely to be initiated where trees are starved or receive insufficient moisture. This makes them unable to carry a heavy crop which in turn stimulates the production of excessive flower buds for the following year.

iii. *Hormonal imbalance:* Extreme alternation in the physiology of trees due to imbalance between growth promoters and growth inhibitors. Among plant hormones, gibberellins (GA_3) have been recognized as inhibitors of flower bud differentiation in several fruit trees. It is also considered that GA_3 from apple seeds diffuse to the shoot where they inhibit flowering. Auxins may also be involved in the fruit induced inhibition of flowering.

iv. *Depletion of carbohydrates and mineral nutrients:* The reproductive effort during the 'on' year leads to a serious depletion of carbohydrate reserves and mineral nutrients. Carbohydrate starvation involves dramatic changes in gene expression. Fruit set and abscission are closely dependent upon carbohydrate levels.

Control

Biennial bearing is difficult to alter or correct. However, it is possible to induce a return to normal yearly fruit production by adopting following practices.

a. *Thinning of fruit buds:* This can be done in early spring before an expected heavy crop year by rubbing off half to three-quarters of the fruit buds, leaving just one or two per spur with the help of thumb and forefinger. The aim of thinning flower buds is to encourage the tree to produce a moderate crop, leaving enough resources for the formation of fruit buds for the following year

b. *Thinning of fruits:* This practice in one year often increases the return bloom in the following year. In a heavy cropping year, thinning is particularly essential for return bloom as the trees become exhausted due to heavy crop.

c. *Watering:* Ensure trees adequate moisture by clearing away competing grass and weeds from around the base of the tree.

d. *Feeding:* Nitrogen application early in the 'off' year has a slight tendency to promote annual bearing in trees having low-nitrogen level.

e. *Pruning:* Judicious pruning of spurs in winter preceeding 'on year' reduces crop load.

f. *Growth retardants:* Application of growth retardant like SADH @ 2000 ppm or Paclobutrazol @ 1500 ppm in June-July before the flower bud differentiation takes place in an 'on year' helps in controlling the problem.

❑❑❑

Apricot

Scientific Name : *Prunus armeniaca* L.

Family : Rosaceae

Apricot is commonly known as Armenian plum. Its scientific name *Prunus armeniaca* derives from the assumption that apricot was known in Armenia during ancient times and has been cultivated there for so long that it is often thought to be native of that place. The cultivated apricot has its origin in North-Eastern China, whereas wild apricot, popularly known as 'Zardalu', appears to be indigenous to India. Generally, there are two types of apricot, namely, sweet kernel type and bitter kernel type.

Apricot is regarded as a nutritious fruit which enjoys world-wide popularity. It is a stone fruit with a nut inside. Fresh apricots are an excellent source of vitamins (A, C and E), carbohydrates, proteins and minerals (phosphorous, potassium and iron). The slightly tart fruit is very versatile and can be used in a number of culinary ways as well as eaten fresh. Besides its use as dessert it is also canned and dried. Apart from this, fruit is processed into jam, nectar and squash. The sweet kernels are used as a cheap substitute for almonds in pastries and confectionery. The bitter kernels are used for oil extraction. The apricot also has numerous health benefits and curative properties.

In India, apricot is grown commercially in the hills of Himachal Pradesh, Jammu and Kashmir, Uttar Pradesh and to a limited extent in north eastern hills. Due to its perishable nature and very short storage life, apricots are marketed in the local markets and nearby cities. Since apricots are very perishable, due care is required before harvesting, during harvesting, packing, transportation and storage. In apricot two physiological disorders are found which are related with high and low temperature resulting in economic losses to the growers. These are discussed below :

1. Pit burn

Pit burn occurs when the apricots are exposed to temperatures above 38°C before harvest. Under such circumstances the flesh tissue around the stone softens and turns brown. This discoloration is due to heat injury and it increases with higher temperatures and longer durations of exposure to sun. This disorder can be controlled by avoiding heat injury by providing some shade to the fruits.

2. Gel breakdown or chilling injury

Gel breakdown in apricot is a physiological problem which is characterized by the breakdown of tissue. This is sometimes accompanied by sponginess and gel formation. In the initial stages there is occurrence of water-soaked areas that subsequently turn brown. This disorder occurs when the fruits are stored at low temperature between 2.2-7.6°C. Such type of fruits affected by chilling injury has short market life and loose flavor due to which it has very less degree of customer acceptability.

❑❑❑

Chapter-5

Avocado

Scientific Name : *Persea americana* Mill.

Family : Lauraceae

Avocado is a pear-shaped berry, which is otherwise known as alligator pear or butter fruit. The term alligator pear is due to the scaly appearance of some varieties of avocado, especially, the Hass avocados, which possesses black leathery skin. The edible portion of avocado is its flesh, which is yellowish green in colour. This flesh has a slightly nutty flavour with the consistency of butter. Avocado contains vitamins A, C, and B complex that sustain accurate performance of neurotransmitter which makes it as a superior food for the nervous system. Avocados have the highest energy value (245 cal/100 g) of any fruit besides being a reservoir of several vitamins and minerals (calcium and magnesium) that can calm and aid restorative sleep which is essential for nervous system to perform efficiently. Avocado is mainly used fresh, in sandwich filling or in salads. It can also be used in ice creams and milk shakes and the pulp may be preserved by freezing. The end quality of avocado determines its nutritional value.

In India, avocado was introduced from Sri Lanka in the early part of the twentieth century. Climatically, it is grown in tropical or semi tropical areas experiencing some rainfall in summer, and in humid, subtropical summer rainfall areas. It is not a commercial fruit crop in India and is grown in a very limited scale in Tamil Nadu, Kerala, Maharashtra, Karnataka and in Sikkim. It is not grown in North India as it cannot tolerate the extreme hot dry winds in summer and frosts in winter. Good potential avocado fruit quality can only be achieved by limiting pre-harvest stress, particularly water stress during the first three months after fruit set, and controlling post-harvest environmental conditions. If these stresses will not be checked, they will result in some physiological disorders in avocado such as leaf burn (pre-harvest disorder) and mesocarp discolouration and grey pulp (post- harvest disorders).

1. Leaf burn of avocado

It is caused by sodium or chloride accumulation in the leaf or by inadequate water supply. The severest injury is observed where irrigation waters are high in chloride.

a) Chloride or tip burn

The most frequent type of scorch in avocado leaves is commonly described as tip burn. The scorch starts at the tip of the leaf and progresses down the blade and sometimes along the margins. This result in decreased functional leaf area of the individual leaf and an even more pronounced decrease in the leaf area of the tree because the severely affected leaves drop prematurely. Occurrence of tip burn is prevalent in spring. Due to chloride toxicity, the affected leaves show necrosis of the tips and the edges and fall, resulting in decline in yield.

b) Sodium scorch

A second leaf-burn pattern caused by sodium accumulation do not start at the tip of the leaf but usually occurs as necrotic or scorched spots near the margin or in the interior area of the leaf. This type of leaf burn sometimes occur singly on the leaf but more frequently, it is found in conjunction with the tip burn due to chloride injury and for that reason has seldom been differentiated from it.

Cause

Heavy irrigation before the end of winter, deep ploughing and incorporation of organic matter into the soil alleviate the problem.

Control

To check leaf burn, avoid heavy irrigation.

2. Pulp spot

Pulp spot is characterized by the blackening of the region surrounding the cut vascular bundles of the fruit- stalk. This is usually localized in nature and the symptom are more prevalent in early season. It is found that the enzyme Phenylalanine ammonia lyase (PAL) activity is higher in pulp spot-affected fruits than in healthy fruits which might be the possible reason for this malady.

3. Mesocarp discolouration or chilling injury

The disorder results in an overall grey to brown flesh discolouration, usually more intense in the distal half of the fruit. The incidence is more prominent towards the end of the season. Fruits that are not cold stored do not show mesocarp discolouration. The fruit which are cold stored for about 28 days show mesocarp discolouration.

About 20 per cent of mesocarp discolouration occurs in initial steps and marginal decline in the percentage is noted during mid-season. However, the incidence of this disorder rises rapidly toward the end of the season showing the marked effect on advanced maturity on mesocarp discolouration in cold-stored fruit. Chilling injury symptoms in avocado, expressed as mesocarp discolouration are found to be associated with embryo growth and ethylene production during cold storage.

4. Grey pulp

This post harvest disorder is generally prevalent in the fruits grown in warmer areas of the country. The main problem is due to late marketing. The incidence of grey pulp develops as the moisture content fall below 80 per cent. Fruits become completely unfit for storage when the moisture content approaches 70 per cent. The extended storage period often results in poor fruit quality due to this physiological disorder.

❑❑❑

Bael

Scientific Name : *Aegle marmelos* (L.) Correa.

Family : Rutaceae

Bael is commonly known as Bengal quince, Indian quince, golden apple, holy fruit, stone apple and bel. It is more prized for its medicinal virtues than its edible quality. The deciduous tree with trifoliate aromatic leaves is traditionally used as sacred offering to 'Lord Shiva'. Fruit is a hard-shelled berry and very well-known for its medicinal properties due to marmelosin content. The ripe fruit is tonic, restorative, laxative and good for heart and brain. The pulp is often processed as nectar or squash. A popular drink called "Sherbet" in India is made by beating the seeded pulp together with milk and sugar. A beverage is also made by combining bael fruit pulp with that of tamarind. These drinks are consumed perhaps less as food or refreshment than for their medicinal effects. For medicinal use, the young fruits, while still tender, are commonly sliced horizontally and sun-dried and sold in local markets. The fruit, roots and leaves have antibiotic activity. The root, leaves and bark are used in treating snakebites. It has 1.8 % protein, 61.5 % moisture, 0.3 % fat, 1.7 % minerals, 2.9 % fiber and 31.8 % carbohydrates per 100 grams of edible part. The fruit is also rich in phosphorus, calcium, carotene, iron, thiamin, riboflavin, niacin and vitamin C. Bael is cultivated throughout India, mainly in temple gardens because of its status as a sacred tree. Bael is a climacteric fruit, taking about 11 months to ripen on the tree. Fruit drop and cracking in bael before ripening are main physiological problems causing considerable losses of the produce.

1. Fruit cracking

Fruit cracking is the physiological disorder in some genotypes of Bael which occurs just before ripening. It is associated with sudden changes in weather conditions such as temperature and humidity. Heavy irrigation or rainfall after a prolonged drought is also one of the causes of this malady.

Control

Cracking can be minimized by maintaining proper moisture up to full growth or maturity of fruit. Mulching also minimizes fruit cracking as it helps in the conservation of moisture fruit cracking can be checked by spraying 0.1 per cent Borax.

2. Fruit drop

The large number of flowers and fruits borne by trees are not carried to maturity and a large portion of this drop occurs during the course of development. Fruit drop is a serious problem in bael which occurs due to many factors such as strong winds, imbalance of soil moisture, improper nutrition and hormonal imbalance.

However, it can be checked to a reasonable extent by maintaining appropriate soil moisture level during fruit development, planting of wind breaks, applying additives to trees and spraying growth regulators like 2,4-D, GA_3 and 2,4,5-T with various concentrations.

3. Chilling injury

Both temperature and duration of exposure of fruit to low temperature during storage are involved in the development of chilling injury. The fruits look sound when kept at low temperature, however, symptoms of chilling injury become evident in a short time after they are removed to warmer temperatures. In chilling injury, an appearance of brown spots on the fruit surface develops during storage of fruits below 9°C.

Control

Storing of the fruits at safe temperature can check this injury.

❑❑❑

Chapter-7

Banana

Scientific Name : *Musa acuminata*

Family : Musaceae

Chilling injury

The word banana is derived from the Arabic word "banan," which means finger and it is known as "the fruit of the wise". In the middle ages, the banana was thought to be the *"forbidden fruit of paradise"*. Bananas are the most popular fruit in the world. Members of the genus *Musa* are considered to be derived from the wild species *Musa acuminata* (AA) and *Musa balbisiana* (BB). It is not a tree, but a perennial herb the leaves of which are usually grouped together, forming a trunk-like structure called pseudostem. The origin of bananas is believed to be Malaysia or India. Bananas are a very important commodity for developing countries, with a clear dual nature as they are at the same time a major staple commodity and a fundamental export commodity. There are almost 1000 varieties of bananas in the world with lot of commercially recommended cultivars. The most commonly known banana is the *Cavendish* variety, which is usually produced for export markets.

Bananas have very beneficial nutritional properties as it contains sugars (glucose, fructose and sucrose) and fiber, which makes them ideal for an immediate and slightly prolonged source of energy. They are a good source of vitamin A, B_6 and C. They are also rich in minerals like potassium, magnesium and iron. The presence of physiological disorders in banana plantations is lowering its economic value as these disorders can lower the quality and market value of fruit or sometimes lead to total loss of the crop. The following are some of the important physiological disorders of banana, cooking banana and plantain.

1. Neer Vazhai

In banana, Neer Vazhai is a serious malady of unknown etiology which affects Nendran and Poovan cultivars in Tamil Nadu (India). The poor plant growth, delayed shooting, lanky bunches with few hands and immature unfilled fingers are the main features of the infested plants. The name 'Neer' means 'water' and 'Vazhai' means banana which has been obtained due to the reason that when the affected fruit is cut, a watery fluid oozes out from it. Although the exact cause of this disorder is not known but it is of serious concern as it results in considerable losses. Severe root damage is noted in the infested plants and it is transmitted through suckers, thus, it can be suspected to be caused by virus or mycoplasma like organisms (MLO's).

Application of growth hormone NAA improves the finger filling, avoiding the extent of this malady.

2. Kotta Vazhai

The etiology of this malady which particularly affects Poovan banana is also unknown. "Kottai" means seed, referring to conspicuously enlarged ovules and immature dark green fruits. The cause of this malady is also not known, however, it is suspected to be associated with incidence of banana streak virus.

The application of 2,4-D (20 ppm) helps in getting normal sized fruits with proper bunch development in the plants, thus anciding Kotta Vazhai.

3. Degrain

This disorder is due to rotting of pedicels which results in drooping of ripe fruits from the bunch. In the mature fingers, high N: P ratio tends to thicken the stalks and flatten the central cylinder. This increases the susceptibility to tannin oxidation to great extent and ultimately promotes rotting of the pedicels of the fingers.

To keep a control on degrain, excess application of nitrogenous fertilizers should be avoided.

4. Peel splitting

Peel splitting is a physiological disorder that occurs in banana, cooking banana and plantain. Peel splitting could results in loss of market quality and value of fruits. It may occur as a result of ripening of fruits at high temperature in a saturated atmosphere, conditions which may develop within polyethylene packaging. It is sometimes observed in the field during bunch development when plants are starved of water. Fruits near full maturity are most susceptible to peel splitting during a dry spell which is interspersed with heavy rains. It may also occur in fruits which have been ripened prematurely. During ripening, the peel looses water by transpiration (to the atmosphere) and by osmosis to the pulp. This results in the increase in volume of the pulp which presumably causes the peel to split. Cultivars/hybrids which have thin peel are more susceptible to peel splitting. Symptoms of peel splitting are characterized by a longitudinal split of the peel, usually beginning from the proximal end near the pedicel. The split usually divides the peel into unequal halves and ultimately exposes the pulp as the split widens.

Control

Following control measures are useful for to check peel splitting.

a) Irrigate field regularly so as to maintain soil moisture.

b) Plant wind breaks around the plantation.

c) Apply some mulch during summer months to conserve soil moisture.

d) Spray 0.8 per cent Borax to check peel splitting.

5. Chilling injury

Chilling injury is the permanent or irreversible physiological damage to fruit tissues, which results from the exposure to chilling temperatures. Even a few hours of chilling temperatures can be sufficient to induce permanent damage. Chilling injury can occur in either unripe or ripe fruit. It can cause lowering of market quality and value or total loss. Symptoms of chilling injury are not readily apparent at the injurious chilling temperature, however, they become increasingly apparent after transfer of fruits to non-chilling temperatures. There are number of commonly occurring visual symptoms which are characteristic of chilling injury in banana such as surface lesions, pitting, large sunken areas, discolouration of the surface and dark water-soaked areas of the peel.

Causes

Several factors contribute to the development of chilling injury in banana. They include

i. Temperature of storage.

ii. Duration of exposure of fruits to the chilling temperature.

iii. Whether exposure of fruits (i.e. cultivar/hybrid) to a chilling temperature is continuous or intermittent.

iv. Relative humidity, composition of the storage atmosphere and post-harvest treatment.

v. Physiological age, maturity or condition of the fruit exposed.

vi. Relative responsiveness (or sensitivity) of the fruit (i.e. cultivar/hybrid) to chilling.

vii. Cultivar and growing conditions.

Control

Chilling injury can easily be avoided in *Musa* cultivars/hybrids by simply limiting storage or handling the temperatures above the critical threshold.

6. Choke throat

It is also known as choke or choking and it occurs when the bunch is about to emerge from the top of the pseudostem but becomes trapped at various stages of emergence. In its most severe form, the bunch fails to emerge from the top of the pseudostem and instead bursts through the side of the pseudostem. Such bunches are worthless and in less severe cases the top 1 or 2 hands become trapped in the throat of the plant leading to badly misshapen fruit.

Causes

Choke throat is seasonal in nature as it is usually worst in the winter and early spring following cold weather. However, it can also occur following periods of water logging or severe water stress and following wind storms. Two factors contribute to the actual difficulty in bunch emergence. Firstly, there is a reduction in the elongation of internodes of the true stem bearing the bunch inside the pseudostem. Secondly, the stiffness of the leaf bases at the top of the pseudostem can prevent proper bunch emergence. Some varieties are more susceptible to this disorder than others. Choke throat is often associated with dwarf varieties.

Control

This disorder can be controlled by following methods.

a. Select taller varieties, which are less susceptible to choke throat.

b. Choose a warm environment, one which is well protected from frosts and strong winds.

c. Control time of bunching to avoid cold weather prior to bunching. Plants bunching in the late spring to mid autumn are less affected.

d. Good on-farm drainage measures including mounding of rows are helpful.

e. Regular irrigation to avoid water stress particularly during hot-dry weather controls the malady.

f. Higher nitrogen rates are thought to be beneficial.

❑❑❑

Chapter-8

Ber

Scientific Name : *Ziziphus mauritiana* Lamk.

Family : Rhamnaceae

Ber is most commonly found in the tropical and sub-tropical regions of the world. Originally native to India, it is now widely naturalized in tropical region from Africa to Afghanistan, China, Malaysia, Australia and in some Pacific regions. The major production regions for Indian jujube are the arid and semi arid regions of India. It is also known as 'King of Arid Fruits' and 'Poor Man's Fruit'. In Philippines it is called *manzana* or *manzanita* ("apple" or "little apple"). Fully ripe fruits are less crisp and somewhat mealy; overripe fruits are wrinkled, the flesh buff-coloured, soft, spongy and musky.

The fruit is eaten raw or pickled or used in beverages. It is quite nutritious and rich in vitamin C, second only to guava and much higher than citrus or apple. In India, the ripe fruits are mostly consumed raw. Slightly under ripe fruits are candied by a process of pricking, immersing in a salt solution. Ripe fruits are preserved by sun-drying and a powder is prepared for out-of-season purposes. It contains 20 to 30 per cent sugar, upto 2.5 per cent protein and 12.8 per cent carbohydrates. The Indian jujube is one of the several trees grown in India as a host for the lac insect, *Kerria lacca,* which sucks the juice from the leaves and encrusts them with an orange-red resinous substance. In Burma, the fruit is used in dyeing silk.

The ber tree is remarkable in its ability to tolerate water-logging as well as drought. Being hardy in nature, it is not confronted with many problems. However, fruit drop is a major physiological problem in ber.

1. Fruit drop

Fruit drop is a major and serious problem in ber production. Generally, the initial fruit set is very high, but the extent of fruit retention varies according to the cultivar, type and on the level of production of endogenous plant hormones. Water-stress results in immature fruit drop. Early maturing cultivars are more resistant to fruit drop while the late maturing cultivars are the most susceptible to fruit drop.

Control

Application of 20-30 ppm NAA, once in the second week of October and again in the second week of November helps in checking fruit drop to a great extent.

2. Chilling injury

Fresh ber fruits are susceptible to chilling injury with development of sheet pitting, i.e. relatively large sunken areas on the surface, if kept at temperatures below 3°C (38°F) for longer than two weeks. Storing ber at safe temperature checks chilling injury.

❑❑❑

Breadfruit

Scientific Name : *Artocarpus altilis*

Family : Moraceae

The breadfruit is believed to be native to New Guinea. It is grown throughout the tropics. Breadfruit is a staple food in many tropical regions. Raw breadfruit has a very light fragrant taste that is somewhat like a guava, but not as sweet. The texture of raw breadfruit is rather tough and rubbery. It is used more as a vegetable than as a fruit. It can be eaten ripe as a fruit or under-ripe as a vegetable. Soft or overripe breadfruit is best for making chips and biscuits. The dried fruit can be made into flour. Its flour is much richer than wheat flour in lysine and other essential amino acids. Breadfruit is high in calories and contains a good amount of vitamin C and potassium with smaller amounts of calcium, iron and magnesium. The wood is yellowish or yellow grey with dark markings or orange speckles; light in weight; not very hard but strong, elastic and termite resistant and is used for construction and furniture. Fiber from the bark is highly durable and is used for clothing in Malaysia.

Hot, humid, tropical lowland is suitable for its cultivation. In southern India, it is cultivated at sea level and on humid slopes to an altitude of 3,500 ft. It is widely grown in the state of Kerala and coastal

Karnataka especially on the sides of Mangalore. Here normal production is 150 to 200 fruits annually. Productivity varies between wet and dry areas. Good quality breadfruits are mature-green, firm, with intact stem and free from defects such as blemishes, sunscald, cracking, bruising and decay. The important disorder of breadfruit is chilling injury which occurs during storage of fruits.

1. Chilling injury

Chilling injury occurs due to storage of breadfruit at temperatures below 12°C (54°F). This type of low-temperature injury becomes more severe as the temperature decreases and the length of exposure increases. Symptoms include a brown discolouration of the skin, pulp browning and off-flavour development and increased susceptibility to decay. Fruit texture will also be adversely affected as the flesh will not soften uniformly. Damage may occur within 7 days of storage at 4°C (40°F), therefore, storing breadfruit at temperature above 12°C can escape the incidence of chilling injury.

❑❑❑

Chapter-10

Carambola

Scientific Name : *Averrhoa carambola* L.

Family : Oxalidaceae

Carambola or starfruit is native to Indonesia. The fruit has ridges running down its sides (usually five) which in cross-section resembles a star, hence its also known as starfruit. It has a waxy, golden yellow to green coloured skin with a complicated flavour combination that includes plums, pineapples and lemons. Carambola consists of two main types of varieties viz. sour and sweet. The sour types contain as much as 1 per cent acid and the sweet types have low acid (0.4 per cent) content with 5 per cent sugars. The acidic nature of the pulp is due to its oxalic acid content. Fruit is a good source of vitamin A, B and C together with valuable minerals like iron. Its most important character is that it is a cholesterol free fruit. Fruits of sweet type are eaten fresh and sour ones can be used for making refreshing drinks, pickling or a substitute for tamarind. Good quality squash, jelly, preserves and candy can also be prepared from carambola fruits. Fruits of sweet varieties can also be dried. The pulp of immature fruits is used for cleaning brassware. Ripe carambolas are eaten out-of-hand, sliced and served in salads or used as garnish on avocado or seafood.

The carambola is a tropical and subtropical fruit. In India, it is called *Kamrakh,* and can be grown upto an elevation of 1,200 m. The fruit suffers from some physiological disorders. The major problem is physical injury, especially on the rib edges that leads to browning. Injury due to abrasion and impact can be avoided by careful handling. Browning due to mechanical injury can intensify with water loss. Fruit that have lost about 5 per cent of their weight due to water loss show visible symptoms of dehydration.

1. Chilling injury

The chilling injury symptoms include surface pitting and rib-edge browning. These pits are either small, less than 1 mm, deep and dark brown or large about 1-2 mm, superficial and light brown. The symptoms occur after 2 weeks at 0°C (32°F) or 6 weeks at 5°C (41°F) storage followed by 2 days at 20°C (68°F). The incidence can be avoided by storing the fruits in safe limit of temperature.

2. Shrivelling

Shrivelling symptoms become visible when the carambolas lose about 5 per cent or greater of their weight due to water stress. The fruit gets reduced in size and is not acceptable to the consumer. Avoid water stress to prevent shrivelling.

3. Physical injury

Surface abrasions and other types of bruising can result from rib edge browning and stem-end browning. The browning intensity increases with water loss from the fruits. Such physical injuries can be avoided by handling carambolas with care to minimize bruising.

❑❑❑

Chapter-11

Cashew nut

Scientific Name : *Anacardium occidentale*

Family : Anacardiaceae

The English name of cashewnut derives from the Portuguese name for the fruit of the cashew tree, *Caju*. It is native to the coastal areas of north-eastern Brazil. It was introduced in Goa by the Portuguese in the 16th century mainly to control soil erosion and to use in afforestation programmes but now it has become an important cash crop of Goa. The fruit of the cashew tree is an oval or pear-shaped accessory fruit known as '*cashew apple*' sometimes called a pseudocarp or false fruit which develops from the receptacle of the cashew flower. The cashew nut is a popular snack with a rich flavour that is often eaten on its own, lightly salted or sugared. Cashew nuts are also sold covered in chocolate. In Goa, the cashew apple is mashed and juice is extracted and kept for fermentation to prepare beverage called fenny.

Cashew nuts are also a good source of vitamin B. They contain useful amount of minerals like potassium, magnesium, phosphorous, selenium and copper. The fats and oils in cashew nuts are 54 per cent monounsaturated fat, 18 per cent polyunsaturated fat and 16 per cent saturated fat. Wide differences in the protein content of cashewnuts (ranging from 13.13 to 25.03 per cent) have been reported from various regions of India. The cashew shell contains 25 per cent of reddish brown oil, industrially known as Cashew Nut Shell Liquid (CNSL) which is a by-product of the roasting process.

In India, Goa has more land under cashew than any other state. Harvesting of cashew in Goa is done generally from March to May. Fully matured apples are allowed to drop down naturally with the attached nut. Then the nuts are seperated from the apples. Nuts are sundried for 1-2 days and stored before selling off in the market. Systematic plant protection schedule should be followed to get steady returns from cashew plantations. Although, the magnitude of the physiological disorders is rather comparatively low in cashewnut, however, yellow leaf spot and little leaf are common problems.

1. Yellow leaf spot

The occurrence of yellow leaf spot, an enigmatic disorder in cashew, is found to be associated with low soil pH (4.5–5.0). The appearance of yellow leaf spot might be due to molybdenum deficiency. The affected leaves are deficient in molybdenum and had excessive amounts of manganese.

The trees sprayed with 0.03 per cent solution of ammonium molybdate once during pre-monsoon (June) and another after end of monsoon (September) is found to be effective in reviving the plants. The disorder can also be controlled by correcting the soil pH with lime.

2. Little leaf of cashew

Little leaf of cashew is caused due to the deficiency of zinc. The leaves remain small in size giving rosette appearance. Application of zinc sulphate (0.5 %) to the soil or by foliar means is helpful in preventing this disorder.

❑❑❑

Chapter-12

Cherry

Scientific Name : *Prunus avium* L.

Family : Rosaceae

Prunus avium means "Bird Cherry" in the Latin language, is commonly known as 'Sweet cherry'. Cherries have pleased the palates of food lovers for centuries. Their ruby-red colour and tangy taste won the hearts of kings, queens and noblemen in the history, making a special place for cherries on their tables. The ultimate celebration of cherries is the National Cherry Festival. It is held every year in July in the "Cherry Capital of the World"-Traverse City, Michigan. This festival originated from a spring ceremony known as the "Blessing of the Blossoms." Thousands of visitors come from all over the world to celebrate the harvest and of course, eat cherries. A delicious fruit, cherry is rich in protein and sugars. Cherry is an excellent source of minerals like potassium, manganese, magnesium, iron, phosphorous and copper. Good amounts of vitamins (A, B, C and K) are found in cherries. They also contain anthocyanins, melatonin and fibers. It has more calorific value than apple.

Sweet cherry was introduced from Europe before India's independence in 1947. Cherry is confined to Kashmir, Himachal Pradesh and hills of Uttar Pradesh in India. Due to higher return, cherry is gaining popularity in temperate regions of the country. However, it is confronted with a serious problem of fruit cracking caused by rainfall prior to harvest and is a limiting factor in the sweet cherry production nearly worldwide. It requires dry climate at the time of fruit ripening. Other minor physiological disorders are pitting, bruising and stem browning. These poor quality affected cherry fruits lose their value for fresh market.

1. Fruit cracking

Fruit cracking is a serious disorder of cherry fruits which results in poor fruit quality leading to heavy losses to growers. Cracking of sweet cherry fruit due to rain near harvest is a major source of crop loss in the cherry industry.

The disorder is characterized by a splitting of the outside layer of the cherry skin called the cuticle. Cracking susceptibility varies with varities and the varieties that take up more water from their fruit surface are likely to be more sensitive to cracking. The varieties that have skins with epidermal cells that are more strongly bound together are likely to resist cracking. Bing cherries have a higher incidence of cracking while Van, Sweetheart, Lapins, Rainier and Sam show lower incidence.

The cultivars with a rapid rate of absorption and a low capacity for expansion crack readily while those with a slow rate of absorption and a high capacity for expansion tend to be immune to this disorder.

Sweet cherry fruit cuticle acts as an efficient water barrier against water uptake from the fruit surface but it contains pores that allow some water penetration. Another important property of the sweet cherry fruit cuticle influencing fruit cracking is its rigidity due to its wax-like structure. Irregular water supply to the fruits causes formation of parenchymatous tissue, thereby, resulting in irregular fruit growth. Concurrent with this the waxed cuticle does not stretch and fissures develop and such cuticular fissures would lead to greater water penetration.

Causes

There are many factors causing fruit cracking. These are as follow.

i. Genetically determined susceptibility of grown species and cultivars. There is a generally accepted opinion that hard fleshed sweet cherry cultivars are more susceptible to cracking than soft fleshed ones.

ii. The quantity (volume) of rainfall at one time.

iii. The full quantity and distribution of rainfall during the ripening season of cherries.

iv. The soil type and soil moisture condition of the cherry orchard.

Control

Any treatment that decreases the rate of water absorption or increases the capacity of the fruit tissues to stretch without rupturing reduces the amount of cracking. Several practices have been developed to prevent cherry cracking due to late season rain exposure. These are as follows:

a) Physical removal of water from cherry surface using air sprayers.

b) Construction of physical barrier such as canopy over rows of cherry trees providing a protective cover which prevents rain water from coming in contact with the cherry fruit surface.

c) A spray of $CaCl_2$ @ 300-350 g/100 lt of water applied at weekly intervals before harvest effectively checks fruit cracking in cherry.

d) Spraying of GA_3 @ 2000 ppm, 3 weeks before harvesting reduces the amount of fruit cracking caused by heavy rainfall following drought.

2. Surface pitting and bruising

This is a very common problem in sweet cherries which causes product rejection and requests for price adjustments by consumers. Pitting can occur during picking, handling, transport to the packinghouse, packing or transport to market. Pits are small sunken areas on the fruit surface whereas larger flattened areas are termed as bruises. Symptoms are primarily caused by a mechanical impact or compression which becomes apparent after several days at room temperature or longer at lower temperatures. Pitting is associated with damage to cells near the epidermis which collapse over time.

Bruising is mainly caused by pickers holding fruit when they remove it from the tree. Bruising appears to be associated with injury to cells much below the epidermis. Such maladies increase cherry respiration, ethylene production and susceptibility to decay and damage.

Control

Fruit factors such as high soluble solids concentration, warm temperature, preharvest use of gibberellic acid and high fruit weight have been shown to reduce fruit susceptibility to damage. Post-harvest pitting damage can be reduced by:

a) Preventing fruit from falling on rough surfaces.

b) Keeping drop heights lower than 30 cm for fruit landing on a smooth surface.

c) Keeping water drop heights less than 20 cm in shower type hydrocoolers.

d) Using the slowest possible fruit speeds on cluster cutter machinery.

Bruising can be reduced by training the pickers properly to touch only fruit stems for picking. Packing house should be designed such that they minimize bruising.

3. Stem browning

Stem browning is another potential physiological disorder which results in discolouration of stem. The packing procedures that scrap or injure stems create wounds that ultimately turn brown. Stem browning can be minimized by proper temperature and relative humidity management. In addition to this, use of chlorine dioxide in hydro-cooler water can reduce development of stem browning.

❑❑❑

Chapter-13

Citrus

Scientific Name : *Citrus* spp.

Family : Rutaceae

Citrus fruits are originated in the tropical and sub tropical regions of South East Asia, particularly in India and China. North East India is the native place of many citrus species. Citrus fruit has been cultivated in an ever-widening area since ancient times. Although oranges are the major fruit in the citrus fruits accounting for about 70 per cent of citrus output, the group also includes small citrus fruits (such as tangerines, mandarins, clementines and satsumas), lemons, limes and grapefruits. Citrus fruits are consumed as fresh fruit or utilized for preparing processed citrus products. Approximately one third of total citrus production is utilized for processing. India ranks sixth in the production of citrus fruits in the world. In India, it is primarily grown in Maharashtra, Andhra Pradesh, Punjab, Karnataka, Uttaranchal, Bihar, Orissa, Assam and Gujarat. Citrus

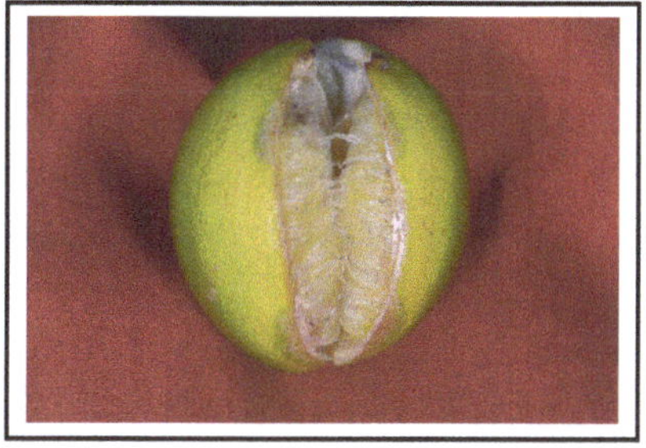

Fruit cracking

fruits have several beneficial health and nutritive properties. They are rich in Vitamin C and folic acid, as well as a good source of fiber. In addition they contain potassium, calcium, folate, thiamine, niacin, vitamin B_6, phosphorus, magnesium and copper. They are cholesterol free.

Trees belonging to genus *Citrus* are small in size and evergreen in nature that is grown in tropical and sub-tropical climates. Citrus fruits are non-climacteric so unlike some other fruits they do not ripen further once removed from the tree, so it is important that they are picked at the right stage of maturity. Lack of proper cultural practices for citrus has tended to restrict exports of this crop. Citrus fruits suffer from a number of different physiological disorders that may considerably affect the fruit crop by destroying the trees and/or the fruits. The major problems include fruit splitting, fruit drop, citrus decline, granulation, etc. Unavailability of proper knowledge against these disorders is the basic reason for low production, short productive life and gradual decline of citrus trees. Citrus production is severely hampered due to these disorders across the world.

1. Fruit cracking or splitting

Cracking or splitting is a common physiological disorder of citrus fruits which is especially related with limes and lemons. Lemons are more prone to fruit cracking than limes. Splitting may be radial (longitudinal) or transverse, radial being more common. The split may be short and shallow or it may be deep and wide, exposing the segments of the juice vesicles. The cracked fruits are breeding ground for micro-organisms making fruit unfit for human consumption.

Causes

The number of fruits affected by this malady varies from year to year. Although the scientists does not assign any exact cause to this malady but number of factors are said to be responsible for this and the most important consideration is that it might be closely related to extreme fluctuations in temperature, humidity, soil moisture and fertilizer levels. It is thought that the trouble is caused by a combination of these factors rather than by a single factor. Splitting is usually observed when growing conditions become erratic such as under water stress, hormonal imbalance, uneven fertilizer supply, temperature fluctuations like hot and cold nights and sudden rainfalls. Mainly fruit cracking is associated with sudden changes in weather conditions, heavy irrigation or rainfall after a prolonged drought. Fruit cracking sometime also result from hot winds. The deficiency of boron and calcium is also found to be responsible for this.

Control

There is no specific remedy to completely overcome fruit splitting, but heavy losses can be reduced by extra attention before and during the critical periods. Reasonable cultural practices should aim to supply optimal growing conditions. Thus, in an orchard, a number of factors should be taken care for controlling fruit splitting. Some of them are discussed below:

a) Supply enough irrigation to ensure a continuous supply of soil moisture. Apply water frequently during summers to avoid drought. Mulching can also be used advantageously for retaining soil moisture by reducing evaporation losses.

b) Use compost and slow release fertilizers to feed the tree. Timed release fertilizers offer the convenience of supplying nutrients at an even rate over the length of the growing season. Use a proprietary mix including organic manure, inorganic fertilizer and biofertilizers which will be most judicious fertilizer application practice.

c) Some varieties are more susceptible to splitting than others, so select cultivars that resist splitting. Young trees or dwarf varieties with relatively small and shallow root systems may be more susceptible to fruit splitting.

d) While planting new trees avoid situations which are more prone to hot desiccating winds.

e) Spray of NAA @ 100 ppm and Borax 0.8% at the period of fruit growth are equally useful in reducing cracking of citrus fruits.

2. Fruit drop

Generally, the citrus trees bear large number of flowers and fruits, however, only few are carried to maturity. Not more than 7-8 per cent of the flowers develop into mature fruits. In spite of very high initial flowering and fruiting in citrus, the ultimate yield is often low due to heavy fruit drop. Physiological fruit drop is a serious obstacle in the development of citrus industry. The fruit drop occurs throughout the fruit development at different times. The shedding of flowers and fruits come more or less in 3 distinct waves.

Pir-head drop

A. *First wave:* The first wave occurs in April soon after the fruit set but before that the unfertilized flowers drop from the trees.

B. *Second wave:* It is commonly known as June drop. It occurs when the fruit is about 3-5 cm in diameter.

C. *Third wave:* Known as pre-harvest drop, this wave reduces the yield considerably. It occurs just before the fruit is matured and this decreases the profit of growers to large extent.

Causes

Following are the causes of the physiological fruit drop.

i. Fluctuation in temperature.

ii. Low atmospheric humidity.

iii. Soil moisture imbalance.

iv. Improper nutrition.

v. Hormonal imbalance (particularly of auxins).

vi. Incidence of insect pest and diseases.

vii. Number of developing fruits.

viii. Number of seed in the developing fruits.

ix. Variety and tree aspect.

The formation of abscission layer due to hormonal imbalance is responsible for flower and fruit drop. The point of attachment with the twig gets weak due to formation of such layer resulting in fruit drop.

Control

Fruit drop can be controlled by maintaining appropriate soil moisture level during fruit development along with the application of 2,4-D (20 ppm).

3. Citrus decline

Citrus occupies an important place of consideration in the economy of the country. In recent years there have been reports from all over country that citrus plantations have started declining in yield and vigour.

Following a period of satisfactory performance, it has generally been observed that trees make excellent growth for first 5-6 years but with increase in cropping, growth slows down and in some cases the trees begin to die back. Affected trees usually do not die instantly but remain in the state of unproductiveness; such trees instead of being source of profit become a great liability to the growers. Citrus decline signifies a continuous dying of twigs, leaves may be small with light green interveinal areas with midrib and lateral veins remaining dark green. In acute cases the growth is usually checked. Shoots have a tendency to die from the growing points downwards. The cropping and quality of such trees also deteriorates seriously.

Causes

Following factors are responsible for citrus decline.

i. *Defective drainage:* As citrus is sensitive to over moist conditions within root zone and thrive best on the soils that are well drained so due to high water table the citrus plantations start declining. High water table was not pre-existing in many citrus growing areas of country. It is only in recent years that it has arisen within 4-5 ft of the surface. Introduction of extensive canal irrigation and network of roads and rails which obstruct natural drainage have been responsible for the rise in water table.

ii. *Poor physiological conditions of the soil:* This malady may be due to poor soil and sub soil conditions as in some areas there are hard pans of $CaCO_3$ or canker layer within 6 feet depth. Due to this hard pan or layer temporarily water logging conditions occur, which are not good for citrus growing.

iii. *Soil erosion:* The root system is exposed to unfavorable conditions due to soil erosion, resulting in poor tree performance.

iv. *Excessive salts:* Salinity is another important factor resulting in injury to citrus plants. When excess boron and $CaCO_3$ are present in soil, it causes deficiency of Zn and Mg. When these salts are present in excessive quantities in soil, tree growth is negatively affected.

v. *Intercropping:* Judicious intercropping do not harm the main trees but farmers grow intercrops which are deep rooted, of long duration and are nutrient depleting. Excessive and indiscriminate intercropping also results in citrus decline and because of the economic reasons owners give citrus plants a secondary position. Ultimately farmers neglect the orchard at the cost of intercrops.

vi. *Nutrient deficiency:* Nutrient deficiency in citrus orchards can lead to serious consequences and ultimately leads to the decline of tree. As farmers do not follow the recommendations with regard to the application of various nutrients, it results in deficiency symptoms of various macro and micro nutrients in initial stages of the plant development and later on show decline. Farmers do not add additional dose of fertilizers for intercrops and ultimately tree shows deficiency and start dying.

vii. *Excessive nutrition:* It has also been observed that excessive dose of nutrient also leads to decline of citrus particularly in case of micronutrients.

viii. *Improper irrigation:* Deficient irrigation and excess irrigation leads to decline in citrus orchard. In some cases water is of poor quality and proves harmful for plant growth. The improper irrigation may lead to decline, because of (a) contact of water with trunks of citrus orchards (b) salinity build up in the soil (c) periodic moisture deficits especially during critical period of fruit setting and fruit growth.

ix. *Vigorous rootstocks:* By the use of vigorous rootstocks like Karna Khatta and Jatti Khatti the trees grow excessively, yields starts declining and quality becomes poor.

x. *Viruses:* Citrus trees have also been destroyed by viruses throughout in India. The viruses results in decline of citrus. The important viruses are greening virus, tristeza, exocortis, psorosis and xyloporosis.

Control

Following practices should be followed starting from planting of citrus crop to control citrus decline.

a. Proper drainage conditions should be provided in the orchard.

b. Maintain physiological conditions of the soil and prevent soil erosion.

c. Avoid intercropping with deep rooted, long duration and nutrient depleting crops.

d. Proper nutrition recommendations for various regions should be followed.

e. Proper irrigation schedules should be prepared and followed to avoid water stress to plants.

f. Avoid insects and diseases by following proper plant protection measures.

Rejuvenation of declining citrus orchard

For rejuvenation of declining citrus orchard, the following schedule is recommended.

- During January-February remove the dead wood before the new growth starts. Apply 2:2:250 Bordeaux mixture immediately and apply Bordeaux paste to the cut surface and the trunk of the trees. Apply Bordeaux paint to the trunk after a week.
- Add recommended doses of farmyard manure and nitrogen to the plants.
- Follow strict spray schedule for the control of insects, pests and diseases.
- Spray the solution containing, 1.5 kg of Zn SO_4 in 500 litres of water and Bordeaux mixture, separately in April, June and September on new growth flushes when the leaves have attained two-third of their size. There should be a minimum gap of one weak between the two sprays.

4. Granulation

Granulation is a complex pre-harvest physiological disorder of citrus fruits which was first reported from California in 1934. This disorder was reported from different citrus growing countries where it was known by different names like 'dry end' in Florida, 'Kaosarn' in Vietnam, 'corkiness' in West Indies and 'crystallization' in California, but 'granulation, is the most widely accepted term for this disorder. A large proportion of fruits are affected by this malady causing huge economic loss to the orchardists, particularly in the arid-irrigated areas of Punjab as well as in the adjoining states of Rajasthan and Haryana. This disorder is characterized by drying up of juice vesicles which become little hard attaining a greyish colour. These vesicles become enlarged with an increase in pectin, lignin and other polysaccharide contents. The granulated juice sacs have excess of Ca, Mg, Na and K and there is considerable decrease in total soluble solids, acidity, juice percentage and sugars in affected fruit. Due to low sugar and acid content, the affected vesicles become tasteless and colourless. The exact cause of granulation is still unknown. However, several factors have been reported to be associated with this disorder. These include frequent irrigation, high temperature during fruit development, high relative humidity, tree age, vigour of plant, large fruit size, high N supply, use of vigorous rootstocks and disturbed auxin concentrations, etc. The factors affecting granulation have been discussed in detail below.

i. *Climate:* Granulation is influenced to varying degrees by weather conditions. Citrus trees growing in humid climates show higher incidence of granulation than trees growing in dry regions i.e. it is favoured by high relative humidity and temperature.

ii. *Age:* Young vigorous trees are more likely to develop granulated fruits than older matured ones.

iii. *Size:* Large fruits are more prone to granulation than small sized fruits.

iv. *Picking season:* Granulation increases with the advancement of picking season.

v. *Species and cultivars:* The incidence of granulation is highly specific to the type of citrus species and cultivars being cultivated. Sweet oranges like Hamlin and Mosambi are more prone to granulation than mandarins. Among mandarins, Dancy tangerine is highly susceptible, whereas Kinnow is free from granulation.

vi. *Rootstocks:* The vigorous root stocks like Rough lemon are found to induce higher incidence of granulation. The highest incidence of granulation was also noted in Blood Red on Jalandhari Khatti, a strain of rough lemon. Blood red budded on galgal and sour orange rootstocks gave the lowest percentage of granulated fruits.

vii. *Mineral nutrition:* The trees having boron and zinc deficiency and high nitrogen content have a greater tendency to produce granulated fruits

viii. *Seed content:* Seed content of fruit has definite relation with granulation. Highly granulated fruits have less number of viable seeds and vice versa.

ix. *Enzymes and plant growth regulators:* An increase in pectin and decrease in pectin esterase activity is associated with granulation. Higher level of auxins, cytokinins and abscissic acid and low level of gibberellins are noted in the granulated fruits than in the normal fruits.

x. *Crop load:* Higher incidence of granulation is recorded in the trees carrying a heavy load.

xi. *Tree location:* Incidence of granulation is more on northern side and in interior of the tree.

Control

No successful method to control granulation has yet been found but certain below mentioned measures help in reducing its incidence and intensity.

a. Early picking of fruits has been recommended before the incidence of granulation becomes serious.

b. Reduced amount and frequency of irrigation reduces granulation.

c. Controlled application of nitrogen fertilizers.

d. Spray of 16 ppm 2,4-D and a mixture of micronutrients like Zn, Cu and K (each @ 0.25%) at monthly interval from August to September is found useful.

e. Spray lime 18-20 kg in 450 lt. of water.

f. Avoid use of sour orange and rough lemon rootstocks which induce vigour and grow resistant varieties.

5. Frenching/Little leaf

It is popularly known as *Foliocellosis* or *mottle leaf* or *little leaf.* It is a typical symptom of Zn deficiency which occurs in all the citrus growing areas of the world and it is the most frequently occurring mineral deficiency in citrus. At the initial stage the parts adjoining the mid-rib and lateral veins remain green. Degree of chlorosis increases with the severity of deficiency when leaves become small, narrow and pointed. The tree gives bushy appearance due to retarded growth. Due to rapid dieback, tree ultimately dies. Zn deficiency can be corrected by foliar spray of 0.5% zinc sulphate.

6. Oil spotting (Oleocellosis)

Oleocellosis is a skin injury caused by oil released from the oil glands in the skin after they have ruptured. Breaking of oil cells due to physical stress on turgid fruits causes release of the oil that damages surrounding tissues. The damage can be caused by physical means (poor handling or abrasion) or climate. Fruit is particularly susceptible to damage during cool wet periods early in the season when the rind is very turgid. The symptoms are greenish or brownish, firm, irregular patches or spots on the skin. Damaged areas are typically darker and become sunken with time. The disorder is sometimes called green spot because the affected areas do not colour at ripening. Oleocellosis can be controlled by avoiding harvesting wet fruit early in the morning or after rain when they are very turgid. Careful handling also reduces severity of this disorder.

7. Chilling injury

The chilling injury is characterized by pitting and brown discolouration. Pits may coalesce to form leathery, brown, sunken areas on the rind. Severity increases with lower temperature below 10°C (50°F) and longer durations of exposure to these temperatures. Waxing or film wrapping can reduce severity of chilling injury symptoms.

8. Frost injury

Frost injury usually occurs after nights with clear skies and little wind causing temperatures below - 2.2°C for more than 4 hours. Damaged fruit may show drier juice sacs and separated segments without injury to the rind. Long term injury includes bark splitting (in young trees) and foliage curling. Frost injury can be prevented through selection of warmer planting sites (elevated ground where cold air can drain away), selection of cold tolerant scion/rootstock combination, wrapping the trunks of young trees, supplying sufficient water to the roots during periods of frost.

9. Albedo breakdown or creasing

Creasing is a major rind disorder of mature navel oranges in particular but other citrus fruits can also be affected. It results in the formation of irregular grooves and furrows in the rind. The problem gets worse on the shaded side of the fruit. Creasing is caused by excessive loss of cohesion between albedo cells (the white layer under the skin) which gets stressed by expansion of the pulp.

Causes

Following factors contribute to creasing.

(i) Rind mineral levels (e.g. low calcium or high magnesium)

(ii) Rootstock (higher incidence on rough lemon)

(iii) Water relations

(iv) Tree age

(v) Nutritional conditions

(vi) Tree health

Creasing can be reduced with pre-harvest sprays of gibberellic acid as well as calcium.

10. Puffing

When fruit are left on the tree for too long time and become over-mature it results in the puffing of fruits in some varieties. It is mainly observed in mandarins, particularly in the easy peeling varieties. The peel separates from the pulp and continues to grow after the pulp growth has ceased. The condition is more prevalent on vigorous trees and can be associated with too much nitrogen. Twigs with luxuriant foliage are more pre-disposed to produce puffy, oversized fruits. As puffing is mainly caused by over-maturity, earlier picking of fruit can minimize the problem.

11. Sunburn

Sunburn of fruit is worse in varieties which produce their fruit on the outside of the trees such as Satsuma mandarins. It can affect citrus trees in various ways. Sunburn is characterized by burning of fruits, leaves and bark which is due to direct sunlight or hot dry winds coupled with inadequate moisture uptake. Sunburn can be minimized by avoiding over-pruning of citrus trees and painting the stems of young trees or exposed limbs with a white, water-based paint.

12. Watermark

Watermark on citrus fruit occurs when a turgid area at the base of the fruit stays water-soaked for a prolonged period in damp, overcast conditions when the fruit is fully coloured. Watermark mainly occurs on Imperial mandarins. Early harvesting as well as treatment with gibberellic acid minimizes the incidence of watermark.

❑❑❑

Coconut

Scientific Name : *Cocos nucifera* L.

Family : Rutaceae

Coconuts are the fruit of the coconut palm, botanically known as *Cocos nucifera,* where *nucifera* means "nut-bearing". The fruit-bearing palms are native to Malaysia and southern Asia. The coconut has been regarded as 'man's most useful tree', 'mankind's greatest provider in the tropics', 'one of the nature's greatest gifts to man', 'tree of life', 'tree of abundance', 'tree of plenty' and the 'tree of heaven' or '*Kalpa Vriksha*' in India. In Sanskrit *Kalpa Vriksha* means "tree which gives all that is necessary for living" because nearly all parts of the tree can be used in some manner or another. The Indian state of Kerala is known as the land of coconuts. The name derives from "Kera" (the coconut tree) and "Alam" (place or earth). It has great antiquity in India which is also evident from the fact that the part of Malabar Coast, extending from South Kanara to Cape Comorin, is known as Keralam meaning the land of coconut. The coconut industry is intimately associated with the rural life and occupation of the people of Malabar Coast.

In Kerala, coconut flowers must be present during a marriage ceremony and many dishes here include coconut. Coconut oil, a saturated fat made from dried coconut meat is used for commercial frying and in candies and margarines as well as in non-edible products such as soaps and cosmetics. In southern Asia, the coconut husk is used in the manufacture of coir, which is subsequently used in the production of rope, as well as household products like door mats and sacks. In India, coconut shells are used as bowls and in the manufacture of various craft products including buttons. In parts of south India, the shell and husk are burned for smoke to repel mosquitoes. The leaves provide materials for baskets and roofing thatch. The white, fleshy part of the seed, the coconut meat, is edible and used fresh or dried in cooking. Coconut water contains sugar, fiber, proteins, antioxidants, vitamins and minerals, and provides an isotonic electrolyte balance, making it a nutritious food source. The sap derived from incising the flower clusters of the coconut is drunk as neera, or fermented to produce palm wine, also known as "toddy" or *tuba* in Philippines and *tuak* in Indonesia and Malaysia.

Coconut produces nuts round the year. Therefore, adequate management of the trees is essential for its unhindered growth. Mismanagement of coconut plantation results in large economic losses. These losses in coconut are mainly as a result of physiological disorders due to poor or inappropriate handling. Of the several maladies that confront coconut production in coconut growing areas, the barren nuts, button shedding and crown choking are of utmost concern.

1. Barren nuts

The occurrence of nuts without or with imperfectly developed kernel is known as 'Barren nuts,' 'Seedless or Imperfect nuts' and it is a common phenomenon in coconut which is as ancient as the cultivation of the coconut. Till about the fourth month, no marked differences are noticeable between a normal and a barren nut. From the fourth month of development onwards it is possible to identify these nuts with little difficulty when they are observed lagging behind in external development. However, in the seventh month, the signs of barrenness become more conspicuous. At this stage, these nuts are generally identified from the normal ones by their peculiar colour, irregular shape and their reduced size and lighter weight.

The barren nuts are characterized by certain conspicuous morphological features such as oblong shape in contrast to round shape of the normal nuts of the same tree. The quantity of husk produced is very much less as compared with normal ones although, the development of husk in these is generally complete. The embryo in the barren nut is mostly absent or when present, it is in varying stages of decay. The decay of the kernel and loss of water inside sometimes occurs due to infestation of embryo by the fungal infection. The most common feature of the barren nut is the frequent splitting of the shell during the period of development. Sometimes the barren nuts show the cracks at the apical end of the nut.

Causes

The phenomenon of barrenness is attributed due to following reasons.

(i) The defective fertilization or poor pollination may be one of the possible causes of barren nut production.

(ii) Nutritional deficiency such as K and B in the palm is also considered to be another cause attributed to the incidence of barrenness.

(iii) Increased production in the palm is also believed by some research workers to be another possible cause for the production of barren nuts.

Control

The production of barren nuts can be reduced by checking the crop load and thinning of nuts. The application of adequate nutrients such as K and B also help to reduce this malady. Application of 1 kg of Muriate of Potash and 200 g of Borax per palm in addition to the regular recommended dose of fertilizer will correct this problem. There is also a popular belief that application of common salt reduces the production of barren nuts.

2. Button shedding and premature nut fall

In coconut, button shedding and premature nut fall are commonly observed disorders which results in considerable loss in nut yield. This phenomenon of shedding of buttons varies from 55 to 95 per cent depending on conditions prevailing and on the variety of the coconut. The coconut is mainly a cross fertilized crop in which pollination is brought about through the agency of wind and insects. A large number of female flowers in the inflorescence consequently fail to fertilize, many of them do not develop into nuts and eventually shed. The female flowers or buttons shed after fertilization and some of the nuts fall prematurely after setting. The fall of immature nuts at a later stage of development often causes considerable loss of crop. It occurs particularly when the fruits are almost fully grown but before the kernel or 'meat' has begun to form.

Causes

The shedding of buttons and nuts in the coconut has been attributed to several widely separated causes, *viz.,* soil and climatic variations, poor soil aeration, nutritional deficiencies, defects in pollination and fertilization, structural defects in the flower, abortion of embryos, physiological condition of the tree at the time of shedding, the limited capacity of the tree to bear fruits, pathological conditions, attack of insect pests, etc. Most of the buttons shed during the first two months after fertilization of flowers. The shedding of buttons is comparatively heavier in the dwarf palms than in the tall ones. The following factors are considered responsible for this problem.

(i) *Imperfect pollination:* The lack of pollination or imperfect pollination is one of the major causes of shedding in the coconuts. Unfertilized buttons shed comparatively sooner. The shedding is attributed to the abortion of carpels in the female flowers and incapacity of the micropylar canal to enlarge due to insufficient nourishment and lack of sufficient moisture in the tissues of the flower, resulting in the failure of the pollen tube to grow and reach the ovule.

(ii) *Weakness of stalk:* Shedding of buttons is also due to loss in mechanical strength of the tissue resulting in the weakness of the bunch stalk. There is formation of abscission layer at the place of attachment to the stalk as a result of severe drought which causes shedding.

(iii) *Moisture imbalance:* The water relationship between the fruit and the plant is of great importance to study the phenomenon of fruit fall. Shedding of buttons not only occur due to moisture deficit but also due to heavy rains. However, maximum shedding of nuts occurs especially after prolonged period of drought and particularly after the onset of the first rains.

(iv) *Soil type:* Shedding of buttons is more severe on heavy soils than lighter soils as heavy soils are more prone to improper aeration.

(v) *Nutritional deficiencies:* In recent years, nutritional deficiency in the soil has been considered to be another factor responsible for the shedding of buttons.

Control

Following control measures help to reduce button shedding and premature fall of nuts.

a) Button shedding is easily managed by irrigation as it increases with period of drought.

b) Ploughing twice in a year results in better soil aeration and controls the problem to some extent.

c) Thinning the bunch after two months by cutting away the nuts that are observed to lag behind in general development and those that are found crowded in the bunch helps to avoid this malady. This practice provides sufficient space for development of the remaining nuts.

d) Application of potassic fertilizers help to reduce button shedding and premature fall of nuts.

e) Root feeding of coconut with 40 ppm of NAA reduces the button shedding and increase the nut yield.

f) Lack of pollination can also be corrected by application of 2,4-D @ 30 ppm one month after opening of the spathe.

3. Crown choking

Crown choking is a serious disorder of coconut characterized by emergence of shorter leaves along with fascinated and crinkled leaves. The leaflets fail to unfurl showing severe tip necrosis. In many cases, it gives a choked appearance to the frond ultimately resulting in the death of the palm. This disorder is mainly associated with the boron deficiency in the coconuts. It can be controlled by application of 200 g of Borax per palm. However, application of 50 g Borax at half-yearly intervals (Feb-Mar and Sept-Oct) along with recommended fertilizer in the basins will control crown choking in the early stages.

❑❑❑

Chapter-15

Custard apple

Scientific Name : *Annona reticulata*

Family : Annonaceae

The custard apple is believed to be a native of the West Indies but it was carried in early times to southern Mexico through Central America. The custard apple, is generally rated as "ugly duckling" species among the prominent members of the genus *Annona.* It is sometimes erroneously termed "sugar apple" or "sweetsop". The custard-apple is also called bullock's heart or bull's heart. The name "Custard Apple" is due to its custard like taste. The fruit has a peculiar aroma which makes a sweet drink and can be used as a milk substitute. Custard apple fruit with a sugary, grainy taste is rich in calories. The fruits are packed with vitamins (A, B and C) and minerals (iron, calcium, magnesium and potassium). The fruit flesh can be scooped from the skin and eaten as is or served with light cream and a sprinkling of sugar. Often it is pressed through a sieve and added to milk shakes, custards or ice cream.

In India the tree is cultivated, especially around Calcutta, and runs wild in many areas where the fruit is eaten only by the poor people. The custard apple tree needs a tropical climate having cooler winters. It does well from the plains up to an elevation of 4,000 ft (1,220 m). It is picked when it has lost all green colour and ripens without splitting so that it is readily sold in local markets. If picked green, it will not colour well and will be of inferior quality. The tree is naturally a fairly heavy bearer. Poor fruit quality and deterioration of fruits due to physiological disorders often results in poor yield. Common disorders observed in custard apples are discussed in this chapter.

1. Stone fruit

In custard apple, some fruits instead of attaining full size, remain very small and become brown. These are known as stone fruits which retain on the tree for a long period even after harvesting normal fruits and also during dormancy and initiation of new growth. Some physiological factors particularly competition among the developing fruits have been suggested to be the cause of stone fruit formation as more stone fruits are observed in the vicinity of developing fruits. The development of such fruits usually takes place in the neglected trees or under moisture stress conditions coupled with nutritional deficiencies. These stresses also force the plants to enter into dormancy well in advance favouring stoning.

Control

Following practices can help in correcting this disorder.

a. Clean cultivation.

b. Adequate manuring and timely irrigation.

c. Better management of plants so that they can flower and their fruits mature in time before the plants enter dormancy without formation of stone fruits.

2. Fruit cracking

Fruit cracking in custard apple may be longitudinal or radial. Sudden and high fluctuations in the water supply to the plants may cause cracking of the fruits. It usually happens from a heavy rainfall or irrigation after a prolonged dry spell.

Appropriate irrigation scheduling can reduce this problem. Evenly distributed irrigation schedule along constant and uniform moisture level in the soil will reduce this malady. Even fruit left on the tree too long will usually crack or split and begin to decay. So harvest the fruits at appropriate maturity to bring cracking of fruits into control.

3. Decline

Tree decline caused by water stagnation is another problem in Custard apple. Such trees shrivel and drying of old branches takes place resulting in sudden tree death. The abrupt decline of custard apple trees may result due to water logging in sandy or rocky soils which also causes root rot. Avoid the stagnation of water in the custard apple orchards. Since water logging causes tree to decline, therefore, heavy soils with poor drainage, sub soils with hard pan or high water table have to be checked.

❑❑❑

Chapter-16

Date palm

Scientific Name : *Phoenix dactylifera*

Family : Arecaceae

Phoenix dactylifera commonly known as the true date palm, is a palm in the genus *Phoenix*, cultivated for its edible sweet fruits. The fruit's English name as well as the Latin species name *dactylifera*, both come from the Greek word for "finger," *Daktulos*, because of the fruit's elongated shape. The date palm is believed to have originated in the lands around the Persian Gulf. Date palms are said to thrive with their *"feet in water and heads in the sun"* because they need plenty of ground water but high heat and arid weather to produce fruit. Therefore, date palms grow best in hot and arid climates.

There are innumerable health benefits of dates. It is an edible, sweet dry fruit rich in proteins, dietary fiber, free from cholesterol and is very low in fat. Dates are very good source of minerals (iron, potassium and calcium) and is rich in vitamins (A, B and C). The juice from the tree (date palm tree sap) is used to make syrup, sugar and vinegar or it may be used as a fresh beverage. After fermentation the sap can be used as a liquor or wine. Seed oil is used in soap and cosmetics. Leaves can be used for making huts and weaving mats, baskets and fans. The trunk may be used for timber for rafters, doors and bridges. The date palm provides protection from wind and cool shelter in hot and dry climate. This tree is ideal as a picnic spot and is also used as fuel. The trimmed fruit stalks are used as brooms. They are also used for making ropes and belts.

Date fruit production is climate sensitive as, if the climate is not warm enough, there will be no fruit production. Incidence of physiological disorders is noted during harvesting, post-harvest handling and marketing. These disorders primarily resulted in postharvest deterioration of quality and quantity of fruit. All these physiological problems are discussed in detail in this chapter.

1. Blacknose

Blacknose disorder is characterized by the abnormal shrivelling and darkening of date tip. The most susceptible varieties are Deglet Noor and Hayani. The pronounced symptoms are excessive breaking of the epidermis, especially in the form of numerous small, transverse breaks at the stylar end of the fruit. The occurrence of shrivelling and darkening is related to humid weather at the Khalaal stage in proportion to the severity of breaks.

Causes

The main causes of blacknose are as follows.

(i) High humidity.

(ii) Rainfall.

(iii) Over thinning.

Control

To control the disorder following practices are helpful.

a) Avoid excessive soil moisture especially at the susceptible stage of fruit development.

b) Avoid over thinning of fruits.

2. Whitenose

In whitenose disorder, the calyx end of the fruit is affected. It is characterized by formation of whitish drying at the calyx end. The disorder results due to dry and prolonged wind in the early Rutab stage causing rapid maturation and desiccation of the fruit. The affected fruit becomes very dry, hard and has high sugar content. This condition can be corrected by hydration in harvested fruits.

3. Crosscuts

The physiological disorder is related to the appearance of crosscuts or V- cuts in the tissues of the fruit stalk bases and on fronds. Fruits with such cuts wither and fail to mature properly. It is an anatomical defect in the fruit stalks and fronds leading to mechanical breaks during elongation of the stalk or the fronds which involves internal, sterile cavities. The varieties having crowded leaf bases are more susceptible to crosscuts and its incidence increases as the palms get older. Khadrawy variety is most susceptible to this disorder.

The disorder can be checked by growing resistant varieties or by reducing the number of fruit stalks in the susceptible varieties.

4. Black scald

Black scald is a minor disorder of unknown etiology. It is characterized by a blackened and sunken area with a definite line of demarcation. The symptoms appear on the tip or the sides of the fruit and affected tissues have a bitter taste. Although, the exact cause is not known, however, exposure to high temperature is considered the main reason behind this disorder.

5. Bastard offshoot

In this disorder the vegetative buds of offshoot fronds show deformed growth. The exact cause is unknown, however, it may be due to reduction in growth caused due to imbalance of growth regulators.

6. Leaf apical drying

The physiological disorder occurs in adult palms due to injury to their root system during transplanting. The symptoms appear on leaves showing apical drying. However, with passage of time, when plants get established properly, the palms recover from these symptoms within two to three years after their transplanting.

7. Barhee disorder

Barhee disorder is characterized by an unusual bending of the crown of Barhee variety mostly to the south and sometimes to the south-west. Sometimes due to increase in severity, the bending reaches at an angle of about 90°. The cause of this disorder is unknown. There is no suitable control measure for this disorder however, it can be corrected to some extent by fixing a heavy iron bar to the opposite side of the bending in order to move the actual weight against the bending side. In this way bending can be corrected within 2 to 3 years.

8. Fertilization injury

Injury due to fertilization affects only young tissue culture-derived palm plants. This occurs during 1-2 years after planting of palms in the field when fertilizers (N, P, K) are applied too close to the palm's stipe. Sometimes its damage is so severe that it could cause the death of the young palm. Avoid fertiliser application too close to the stipe of palm.

9. Frost damage

It is a serious disorder causing injury to date palm leaves showing partial or total desiccation. The protoplasm freezes after coming out from the cells when temperature falls below 0°C. The affected leaves turn brown and desiccated. The severity of damage is related to the intensity and duration of frost. The leaflet ends become yellowish and dry up at -6°C, leaves of external crown desiccate at -12°C and leaves of middle crown freeze from -15°C.

If low temperature gets prolonged then the injury reaches to the central crown resulting in desiccation of whole foliage and the palm seems to be completely burned. It ultimately results in the loss of fruit on the palm. Such loss of leaves from palm does not support the maturity of fruit crop in the next year.

The problem can be checked by irrigating the plants and keeping them wet when temperature begins to fall below 0°C as it stores some heat and give protection.

10. Darkening

High moisture content and temperature results in both enzymatic and non-enzymatic browning of dates results in darkening of fruit. It can be checked by keeping the temperature and moisture content low.

11. Skin separation or puffiness

In puffiness disorder the fruit skin gets dry, hard and brittle, and gets separated from the flesh. This disorder develops during ripening of soft date cultivars, which vary in susceptibility. At a stage before the beginning of ripening, the factors such as high temperature and high humidity may predispose the dates to skin separation. Therefore, controlling high humidity and temperature can check puffiness in dates.

12. Sugar spotting or sugaring

Sugar spotting occurs due to crystallization of sugars below the skin in the flesh of soft date cultivars. Sugaring alters fruit texture and appearance although it does not influence taste. Incidence and severity of sugar spotting increases with the storage temperature and time. This disorder occurs mainly in cultivars in which glucose and fructose are the main sugars.

Sugaring may be reduced by gentle heating of the affected dates. Storage at recommended temperatures minimizes this disorder.

❑❑❑

Durian

Scientific Name : *Durio zibethinus* L.

Family : Malvaceae

Durian, an exotic fruit, native to Malayasia and Indonesia, is considered to be the 'King of all fruits' in South East Asia. The term 'Durian' has been derived from the Malay word 'Duri' which means thorn. The fruit has a tough, thick, thorny husk that resembles the husk of jackfruit. The notorious odor has given rise to the unflattering terms like "Civet Cat Tree", and "Civet Fruit" in India and *"Stinkvrucht"* in Dutch. Durian is delicious, soft and succulent with a distinctive sulphuric smell. It is characterized by the rich, buttery smooth and luscious flesh. The fruit is distinctive for its large size, unique odour and formidable thorn-covered husk. The fruit is a good source of carbohydrates, protein, dietary fiber and energy. Some of the other durian fruit benefits are their high iron content and lack of cholestrol. It is rich in vitamin B, C and E, as well as in amino acids. The jelly is prepared from over ripe fruits which is sour in taste.

Durian is a seasonal fruit which lasts typically from June till August. Its season coincides with that of the other specifically tropical fruits like mangosteen, jack fruit and mango. In India, there are no large orchards or commercial plantings of durian, but cultivation of some trees is found in and around Nilgiris (Tamil Nadu) and West Coast. In India, durian trees are often planted along the banks of streams, where the roots can reach water. Fruits take about three and a half months to reach maturity. Immature fruits are picked only for use as 'vegetables'. The rich aroma develops as fruits ripen and reach their peak in 3-4 days after harvesting and becomes soft. The ripe fruits deteriorate rapidly and cannot be transported over long distances. There are few physiological disorders of durian fruit which are discussed in this chapter.

1. Chilling injury

The black discolouration of surface, especially the groove between thorns is a peculiar symptoms of chilling injury in durian. In such a condition the fruit fails to ripe due to loss of ability to convert starch into sugars. Almost all durian cultivars are damaged by chilling injury due to storage at 5°C (41°F) for one week or 10°C (50°F) for two weeks. However, chilling sensitivity varies according to the cultivars. The incidence can be controlled by storing the fruits at safe temperature.

2. Uneven fruit ripening

A common problem has been observed in durian that fruits ripen unevenly. In such circumstances a portion of the aril remains hard, leathery, whitish in colour, odourless and tasteless. A higher incidence of this disorder may be observed in larger fruits than smaller fruits. Incidence and severity are related to pre-harvest factors. It may be related to a combination of factors such as nutrition, water availability and environmental conditions. There is a close association of imbalances in mineral nutrition with this physiological disorder. To control this disorder apply calcium nitrate at 2 kg per tree one month before harvest.

3. Wet core (water core)

In this disorder, flesh areas appear water-soaked resulting in deterioration. The affected areas get decayed faster than unaffected areas. Water core is caused by rain just before harvesting. The excessive accumulation of water through the roots during maturation time of fruits is another cause behind this problem. The disorder can be corrected by avoiding excessive watering.

4. Flesh burn

Flesh burn is characterized by browning of the flesh. The disorder degrades the quality of fruits which fetch poor price in the market. The main reason behind flesh burn is boron deficiency. It can be controlled by the application of Borax @ 10g per tree every year. The appropriate time for Borax application is during flowering.

❑❑❑

Chapter-18

Fig

Scientific Name : *Ficus carica*

Family : Moraceae

Fig is one of those fruits which are known to mankind since early ages. Legend maintains that the Fig was the "Autumn Fruit' as revealed by the Greek Goddess, Demeter, and the tree is still held sacred in most parts of the Mediterranean. The fig's nutritional value is so high that it is often referred to as the 'Fitness Fruit'. It is a rich source of dietary fiber. It contains a good amount of vitamins and various useful minerals like iron, calcium, magnesium, copper, etc. The matured fruit has a tough peel either pure green, green suffused with brown, brown or purple. The interior is a white containing a seed mass bound with jelly-like flesh.

Fig trees are a popular home plant and are easy to grow in warm climates. Although, the fig grows best and produces the best quality fruit in dryer warm-temperate climates, however, with extra care figs can also be grow in wetter, cooler areas. Rains during fruit development and ripening can cause the fruits to split. Because of losses in transport and short shelf life, figs are a high-value fruits of limited demand. Figs must be allowed to ripen fully on the tree before they are picked. They will not ripen if picked when immature. So, harvest the fruit at optimum maturity and gently to avoid bruising. In the field fig fruit faces certain problems that are caused due to abiotic stressess.

1. Sunburn

Sunburn is a serious problem of newly planted young fig trees. The exposure of plants to direct sunlight and heavy pruning exposing the trunk and branches, is responsible for this disorder. The affected parts crack and the bark peels off. The affected trunk or branches may be a centre for fungal infection. White washing of exposed parts especially stem can prevent this malady.

2. Fruit splitting or cracking

Splitting is attributed to sudden change in atmospheric humidity during ripening. High humidity coupled with low temperature usually results in fruit splitting and low fruit quality. This makes the fruit unfit for consumption as the pulp is exposed to insect and microbial infection. The heavy rains and excessive or sporadic watering may cause the fruit to split. The amount of splitting varies from variety to variety. The riper the figs, the more they will split. These cracked fruits may also result due to nutritional deficiencies, but excessive nitrogenous fertilization also results in splitting. Excessive pollination by fig wasps from a nearby caprifig tree also causes the syconium to split open.

Splitting can be reduced, however, if its fruits are picked just before full maturity and over nitrogen fertilization is avoided. It is also advisable to keep fig trees at a safe distance from the male caprifig to avoid excessive pollination.

3. Fruit drop

One of the most common fig tree problems is fruit drop. The environmental factors like excessive drought and heat, cold nights or light frost results in fruit drop in fig. Drought or inconsistent watering is the most common reason that fig fruit falls off the tree. Apart from this lack of pollination also causes fruit drop in figs. Typically, if there is a lack of pollination, the fig fruit will fall off while it is still very small, as the tree has no reason to grow them larger as they will not produce seeds without proper pollination. Rapid temperature changes, to either very hot or cool can cause fig fruit to fall off the trees. Sometimes insect, pest and disease infestations also results in fig fruit drop.

It can be controlled by frequent watering, proper fertilization and general care of the tree to keep them healthy and prevent disease. Provide adequate protection for a fig tree that may have to go through a rapid temperature change. Spray of GA_3 @ 30 ml/litre of water is effective for controlling fruit drop.

❑❑❑

Chapter-19

Grape

Scientific Name : *Vitis vinifera* L.

Family : Vitaceae

Grape cultivation is one of the most remunerative farming enterprises in India. Considered the *"Food of Gods"* by the ancients, the little juicy ball holds more mysteries than we can think of. Classified by biologists as a 'true' berry, it has fleshy insides all the way to the tiny seed, making it a strong powerhouse of nutrition and energy. As one of nature's richest source of antioxidants, the grape is priceless in its nutritional value. In fact many dieticians believe that it helps lower the incidence of two of the deadliest diseases in the urban world- coronary disease and cancer. In its juice form, the grape is known to cleanse the liver and remove excess uric acid from the body. One interesting but little known fact is that grape seeds are healthier than the fruit itself.

Rotting of berries

They contain powerful antioxidants serving to prevent premature ageing, disease and decay. Grapes, one of the most delicious fruits, are rich sources of vitamins A, C, and B_6. In addition to this they contain essential minerals like potassium, calcium, iron, phosphorus and magnesium.

Grapes are one of the most popular fruits all over the world. Apart from their sweet, juicy taste, the other reason that makes them a favourite among masses is, its simplicity of consumption since, you don't need any cutting or peeling. Although grape cultivation is considered as highly remunerative but some constraints in the form of physiological disorders result in high risk of losing the crop and very low proportion of export quality grapes. All such maladies have been discussed in this chapter.

1. Barrenness

Barrenness of vines is one of the most exasperating problems in India especially in North West plains. This problem is sometimes so serious that the vines are left with few productive canes. The failure of vines to bear normal crop due to development of unproductive wood reducing productive life of vines are major features of barrenness in our country. This usually happens where orchards are neglected and cultural practices are defective. The apparently healthy canes and arms when given cuts are observed to be dry or partially so. The vines do not bear flower bunches and if few bunches are found they are small in size. Lack of flower bud formation or their subsequent death results in lack of flower bunch formation. The cause of death of the floral primordia was reported to be shading. The excessive foliage creates such ecological conditions due to which some saprophytic fungi turn into parasitic leading to the death of tissue. The excessive foliage is due to over fertilization especially with nitrogen.

Barrenness is a common phenomenon in Anab-e-Shahi cultivar of grapes trained on Bower system. Loss of floral primordia was found to be a factor contributing to barrenness in this cultivar. The fruitfulness of a bud varied both with the diameter of the cane and its nodal position. The fruitfulness of canes declined when the cane diameter increased or decreased from 1.2 to 1.4 cm. There is close and highly significant relationship between the vigour of a vine and the mortality of its canes.

The important factors which contribute to the barrenness of vines are defective training and pruning practices, bud failure and inadequate care during the non-bearing period. In north India especially Punjab, barrenness of vines is mainly due to defective training and inadequate trellis support. These vines are allowed to grow very dense and the development of adequate number of bearing canes is not encouraged. Moreover, these vines are not trained to develop primary and secondary arms and fruiting canes. On the other hand, faulty pruning practices in north India and continuous double cropping in south India lead to the barrenness.

Control

The following practices can help in preventing or delaying barrenness of vines.

a) Adoption of proper training and pruning practices by the grape growers.

b) Following adequate plant protection measures against insect-pest and diseases.

c) Training the vines systematically under the guidance of a technical expert by following proper training system.

d) Avoiding the practice of double cropping.

e) Increasing the fruitfulness of buds and bud break by following suitable cultural practices to maintain the productivity of vines.

f) Avoiding heavy yields from the young vines so as to prolong their productive life span.

g) Heading back the barren vines to retain healthy and sound limbs to maintain the productiveness of vines.

2. Water berries

The phenomenon of development of unusual berries in the grape clusters is very common. There is lack of normal sugar, colour, flavour and keeping quality in such berries. These berries look like small cellophane bags, partially filled with sap, which remain hanging on bunches. Although these berries look normal in size but when touched their pulp is very soft instead of firm. In severe cases, the berries are dull in colour, get shrivelled and dry as harvest time approaches. Water berries are generally confined to the tip of main rachis or its branches.

Causes

Viticulturists feel that development of water berries result due to following reasons.

(i) Overcropping and inadequate nourishment of all the berries in a cluster.

(ii) Frequent watering and more application of nitrogenous fertilizers resulting in excessive shoot vigour.

Control

Water berry formation can be reduced by taking following steps.

a) Bunch and berry thinning accompanied by cluster clipping.

b) Checking the shoot growth by limiting water and nitrogen applications.

c) Applying potash and oil cakes.

d) Spraying boric acid (0.2%).

3. Shot berries

It is a major problem in varieties like Beauty Seedless and Perlette which bear compact bunches. Shot berries are usually smaller, round and seedless as compared to the normal berries. In seedless varieties, they are smaller in size and sweeter than the normal berries of a bunch, while in a seeded variety these are seedless and small. Such shot berries are also known as *Millerandage*. In general, small berries shatter after 7-20 days after set due to lack of nourishment or failure in embryo development. On the other hand, the berries which get inadequate nourishment eventually develop into shot berries.

Causes

The causes of development of shot berries are as follow.

(i) Poor pollination or fertilization.

(ii) Poor carbohydrate nutrition to the flowers.

(iii) Compactness of bunches.

(iv) Deficiency of boron.

(v) Inappropriate application of gibberellic acid.

(vi) Girdling at incorrect stages.

Control

a) Apply gibberellic acid (GA_3) or girdling immediately after fruit set and before the berry shatter stage.

b) Avoid boron and zinc deficiencies.

c) Berry thinning also help in avoiding incidence of shot berries.

d) Dipping of bunches at berry set stage in ethephon (25 ppm) + sevin (2000 ppm), also helps in controlling the problem.

4. Hen and chicken disorder

The name 'hen and chicken' is given due to the reason that many shot berries surround a bold berry in a cluster, where bold berry is compared to hen and shot berries to chicken. The characteristic feature of formation of shot berries in a cluster is their number which is several times more than the normal ones. The deficiency of zinc and boron are found responsible for this disorder and in these cases the shot berries are seedless. The shot berries are spherical or sometime oblate i.e. compressed at ends in case of boron deficiency as against elongated shape of normal berries whereas these are reduced in size with normal shape in case of zinc deficiency.

To reduce the incidence of hen and chicken disorder, it is advisable to correct the deficiency of zinc and boron by following proper spray schedule before flowering.

5. Uneven ripening

The presence of green berries in a ripe bunch of coloured grapes is called 'uneven ripening'. It is a varietal character and a common problem in Bangalore Blue, Bangalore Purple, Beauty Seedless and Gulabi grapes. The problem varies from bunch to bunch even within a variety. The major reasons attributed to uneven ripening are inadequate leaf area and non availability of reserves to a developing bunch.

Cultural practices like cluster thinning, girdling and use of growth regulators can reduce uneven ripening. Application of ethephon (250 ppm) at colour break stage (verasion stage) is recommended to reduce this problem.

6. Blossom-end rot

In this rot the development of black sunken spot at the blossom-end of the berry is seen which later on spreads with water soaked region around it. Defective calcium nutrition and assimilation appears to be the cause for it. This disorder can be corrected with a spray of 1.0 per cent calcium nitrate.

7. Interveinal chlorosis

This is a nutritional disorder which may be due to magnesium, zinc or iron deficiency. In interveinal chlorosis area in between the veins becomes yellowish. The deficiency can be corrected by spraying vines with 0.2 per cent sulphate salts of the nutrients. Likewise, under high salinity conditions, marginal chlorosis which is followed by progressive necrosis of leaf blade towards petiole may also occur.

8. Pink berry

It is a common disorder in bunches of Thompson Seedless and its clone Tas-a-Ganesh in Maharashtra. Pink blush develops on few ripe berries close to harvesting. The pink colour turns to dull red colour resulting in unattractive bunches. These berries become soft and watery and do not stand for long after harvesting, so these get deteriorated during storage and transportation.

Causes

Although the definite cause of this disorder is unknown but scientists believe that excessive use of ethylene is a major cause of such berry formation. Apart from this scientific evidences also reveal following factors to be responsible for the incidence of pink berry.

i. Large diurnal variation in temperature during the period of berry ripening.

ii. Application of potassium to the vines prior to berry softening.

iii. Application of calcium sprays after berry softening.

iv. Prolonged soil moisture stress after verasion stage.

Control

The spray mixture of 0.2% ascorbic acid and 0.25% sodium diethyl dithiocarbonate at fortnightly intervals commencing berry softening are useful to control this disorder.

9. Bud, flower and berry drop

As a result of heavy drop of buds, flowers and subsequently berries the bunches become very loose and unattractive. The unattractive berries fetch low prices apart from market reduction in yield. Flower bud and flower drop is a very serious problem in some grape varieties under north Indian conditions. In French it is known as '*Coulure*' or '*Shelling*'. The condition is characterised by excessive shedding of flower buds about 8-10 days before full bloom. Even buds may drop with a slight jerk to the panicle. It is also accompanied by a stage known as '*Shatter*' where the ovaries of flowers that fail to develop into berries drop down from the clusters within a week or so after anthesis. The conditions sometimes gets worse leading to very loose and straggly bunches as a result of excessive shattering.

The basic reason for '*Coulure*' is the formation of weak abscission layer at the point of pedicel attachment of the rachis. The major contributory factors are moisture stress, C/N ratio and hormonal imbalance with high ABA and low auxin levels. Different varieties respond differently to Coulure. The seriously affected varieties are Beauty Seedless, Thompson Seedless, Cardinal and Gold. Post harvest berry drop is due to weak pedicel attachment to the berries. Berry shattering is common in Anab-e-Shahi, Cheema Sahebi and Beauty Seedless cultivars.

In some cases when panicles are fully expanded, the flower buds drop before the fruit set. The drop including complete panicle drying in some cultivars have been also reported. This menace may be due to improper nitrogen application, carbohydrate nutrition, improper fertilization of berries, heavy crop load, uneven ripening, Zn and B deficiency and auxin deficiency at a particular stage of berry development.

To control bud, flower and berry drop, the following appropriate management practices have been suggested.

a. Exogenous application of plant growth regulators particularly spraying NAA (50 ppm) a week prior to harvesting can minimize post harvest berry drop.

b. Appropriate canopy management practices that allow good ventilation and sunlight to the vines.

c. Girdling before flowering improves fruit set due to increased accumulation and availability of photosynthates to flower panicles. Making 0.5 cm wide girdle on the trunk of vines about 10 days before full bloom reduces bud and flower drop resulting in better set and reducing subsequent berry drop.

d. Avoid prolong water stress prior to flowering. Apply good quality water at frequent intervals and avoid excessive irrigation at bloom.

e. Boron and zinc must be applied to vines well in time.

f. Application of 500 ppm ethrel at ripening stage in Beauty Seedless results in even ripening, reducing the possibility of pre-harvest drop.

g. Regulating yield per vine can also reduce this problem.

h. Thinning practices also reduces berry shattering. On the other hand, flower bud thinning and berry thinning is recommended for improving the size and quality of berries.

10. Berry shrivel

It mainly occurs in seeded grapes and about 5 to 30 per cent of clusters may be affected. The disorder is occasionally so severe that vineyard is removed. The exact cause is unknown but it is thought to be physiological one. The cultivar Emperor is most prone to this malady. The berries become flaccid and sunken as the ripening begins. The flaccid berries are usually interspersed with normal ones. The cluster may be slightly dull green but the berries must be touched to confirm the flaccid condition. The amount of fruit involved usually increases upto harvest. Spray of gibberellins @ 1.6 gm/ acre applied about two weeks after fruit set when berry diameter averages 10-15 mm will reduce the shrivelling by 50 to 70 per cent.

11. Cluster apex wilt

In this problem, the apical portion of the cluster will contain wilted or shrivelled berries and the green fruit at the apex will be soft to touch. The problem becomes evident in Thompson Seedless variety before the onset of ripening. The wilted berries also remain underdeveloped and are sour in taste. This problem is usually more prominent in overbearing clusters and weak shoots. The apex of the cluster dries up resulting in the formation of small and light brown berries in severe cases. This malady is also favoured by insufficient water supply to the developing berries as a result of deficit transpiration on such shoots. Further, this disorder is associated with excessive N supply.

Application of optimum moisture and nitrogen to the vines controls the disorder.

12. Stalk necrosis

The dead areas appear on the pedicel of the flowers or stems of panicles. This disorder results in drying of a portion of clusters or panicles beyond the dead area. The calcium deficient vines are more prone to this malady.

13. Berry cracking and rotting

During the ripening stage, the grapes are highly susceptible to excess water supply. In north India pre-monsoon showers during berry ripening destroy the whole grape crop as the ripening berries get cracked and rot immediately after rains. This is attributed to heavy absorption of water by roots which is then translocated to the ripening berries. This results in the sudden upsurge in the berry contents. Due to dry conditions before rain, the berry peel loses its elasticity and becomes thin. After rains the pulp expands due to the upsurge but peel do not expand due to its inelasticity resulting in cracking. All the varieties of north India are susceptible to this problem but Thompson Seedless, being a late ripening variety suffers a lot from this malady. On the other hand, in south India, Bangalore Blue and Anab-e-Shahi are less prone even if rain occurs during their ripening period. This is due to the reason that Bangalore Blue is slip skin variety and Anab-e-Shahi contains lesser sugar content at ripening.

Control

Following practices can help to escape from this disorder.

a. Early pruning of vineyards to avoid the coincidence of ripening with the normal rainfall.

b. Hastening the bud burst and ripening by some suitable means.

c. Growing early ripening varieties like Pusa Navrang ans Pusa Urvashi.

d. Using Dormex (0.2%) after pruning to break dormancy in mid January.

❑❑❑

Chapter-20

Guava

Scientific Name : *Psidium guajava*

Family : Myrtaceae

Guava is native to tropical America, commonly known as 'Apple of Tropics' and it has got much greater health benefits than apple, therefore, it is also known as 'Poor man's Apple'. It is widely grown in the tropics, enriching the diet of million of people in the world. Guava has spread widely throughout the world because it thrives in a variety of soils, propagated easily and bears fruit relatively quickly in comparison with other fruits. At the beginning of the 16th century the Portuguese introduced it throughout the Pacific as far as the Philippines and the Spanish took it to India. Now, guava is one of the important commercial fruits in India which ranks fourth most important fruit after mango, banana and citrus. Guava contains key nutrients like vitamin C, carotenoids, folate, potassium, fiber, calcium and iron. The vitamin C content in guava is four times more than an orange besides this it also contains moderate quantities of vitamin A. The leaves are particularly rich in flavonoids like quercetin to which most of the medicinal properties of guava are attributed and which are also found to be responsible for the anti-bacterial activity of guava leaves. Quercetin is also responsible for the anti diarrheal action of guava leaves since it relaxes the smooth muscles of the gastrointestinal tract and also inhibits the bowel contractions.

Guava is successfully grown all over India. It is quite hardy, prolific bearer, commercially significant, highly remunerative crop even without much care. However, certain strategies are needed for enhancing guava fruit quality in India in order to be competitive in the world market. This includes adoption of appropriate cultural practices and management of physiological disorders for achieving good quality fruits. Some important disorders found in guava are discussed below.

1. Fruit drop

Fruit drop is of serious concern in guava as it results in about 45-65 per cent fruit loss which is attributed to different physiological and environmental factors. It may be the result of poor soil, inadequate irrigation water, attack of insect pest and diseases, depletion of nutrients, hormonal imbalance, etc. Spraying of GA_3 has been found to be effective in reducing the fruit drop in guava.

2. Sun scald

Guava fruits exposed to direct sun light after being in the shade for an extended period of time gets scalded. The leaves of the tree are also affected by sun scald, particularly on a bright sunny day following a period of warm cloudy humidity. To protect the guava fruits from solar radiation while on the tree, they can be covered with paper bags.

3. Bronzing

Bronzing of guava has been observed in places having low soil fertility and low pH. Affected plants show purple to red specks scattered all over the leaves and under aggravated conditions total defoliation and fruits characterized with brown coloured patterns on the skin with reduced yield are noticed. Foliar application of 0.5 per cent diammonium phosphate and zinc sulphate in combination at weekly intervals for two months reduces the bronzing in guava.

4. Fatio

This disorder is caused by deficiency of organic matter, nitrogen, zinc and boron. Supplementing soil with organic matter, nitrogen, zinc and boron helps in checking fatio.

5. Chilling injury

Symptoms include failure of mature-green or partially ripe guavas to ripen, browning of the flesh and in severe cases the skin also appears brown with increased decay incidence. Severity upon transfer to higher temperatures is also noticed. Fully-ripe guavas are less sensitive to chilling injury than mature-green guavas and may be kept for up to a week at 5°C (41°F) without exhibiting chilling injury symptoms.

❑❑❑

Chapter-21

Hazelnut

Scientific Name : *Corylus avellana* L.

Family : Myrtaceae

Hazelnut is also known as a cob nut or filbert nut according to species. The hazelnut was native to the Black Sea coast long before our era, where it was found growing wild. Turkey is one of the few countries in the world having a favourable climate for hazelnut production. Hazelnuts are extensively used in confectionery to make praline and also used in combination with chocolate for chocolate truffles and products such as Nutella. Hazelnut oil recovered from hazelnuts is strongly flavoured and used as cooking oil. Hazelnuts are a good source of energy with 60.5 per cent fat content. Hazelnuts have been ranked as one of the most nutritious nuts because they supply significant amounts of protein and fiber, vitamin E and B, minerals like iron, zinc, calcium, magnesium and potassium. These have been deemed the most beneficial nut for heart health, containing a mix of beneficial compounds that protect against coronary disease. They are rich source of oleic acid which helps to lower cholesterol. Hazelnuts lower blood pressure because of minerals such as calcium, magnesium and potassium. Apart from nutritive value hazelnut wood is used to make bows.

In India seedling trees are found growing wild in Shimla, Kinnaur and Chamba districts of Himachal Pradesh. In Chamba district of Himachal Pradesh they are known as Thangi. The hazelnuts should be properly sized to meet the stated market type and should be properly filled with at least 50 per cent of the shell cavity occupied by nutmeat. Shells should be free of disorders like kernel black tip, blank nuts etc.

1. Kernel black tips

Black tip on kernels appear to be associated with nuts having split or weak sutures. It appears to be caused by an oxidation process that occurs on the pellicle only. It deteriorates the nut quality due to which it fetches less returns.

2. Blank nuts

In this type of disorder, hazelnuts are devoid of normal kernels which result from defective embryo sacs, unviable eggs and failure of fertilization or embryo abortion at varying stages of development. Poorly filled nuts may also result during an over-productive year resulting due to the alternate bearing cycles.

3. Brown spots in the kernel cavity (BSKC)

It is a commercially important physiological disorder particularly found in Spanish hazelnut cultivars. Several factors are associated with this disorder in the kernel cavity which occur in the last stages of kernel development. BSKC can affect 7 to 97 per cent of the total production, depending on the year. A direct correlation is observed between the percentage of fruit affected and the altitude of the growing area. Although the definite cause of the BSKC disorder is not known, it seems that orchards at high elevation have a lack of accumulated heat units during the period of kernel development which occurs during end of June to mid-July, that is correlated with an increase in the incidence of BSKC. Therefore, it seems that the disorder could be related to some enzymatic process that could be affected by temperature.

❑❑❑

Chapter-22

Jackfruit

Scientific Name : *Artocarpus heterophyllus*

Family : Myrtaceae

Jackfruit is believed to be indigenous to the rain forests of the Western Ghats of India and the name "Jack fruit" is derived from the Portuguese word *Jaca*, which in turn, is derived from the Malayalam language term, *Chakka*. The jackfruit has played a significant role in Indian agriculture for centuries. Archeological findings in India have revealed that jackfruit was cultivated in India 3000 to 6000 years ago. During the season, each tree bears as many as 250 large fruits, supposed to be the largest tree-borne fruit in the world. The fruit varies widely in size, weigh from 3 to 30 kg and has oblong or round shape measuring 10 cm to 60 cm in length and above 25 to 75 cm in diameter. The unripe fruits are green in colour; when ripe, might turn to light brown colour and gives pungent smell. Like durian fruit, its outer surface is covered with blunt thorn like projections which become soft in ripened fruit. Jack-fruit is rich in dietary fiber, vitamins (A, B and C), minerals (potassium, magnesium, manganese and iron) and antioxidant flavonoids like β-carotene and lutein. Unripe jack fruit is used like vegetable in the preparation known as *"Kathal sabzee"* in some north Indian states. The canned product is more attractive than the fresh pulp and is sometimes called "vegetable meat". Maturing in 35-40 years, their wood can be used for furniture. The gum from this tree is used to repair small holes in pots.

In India, the total area under jackfruit cultivation is thought to be very less as it is mainly grown in back yards and as intercrop amongst other commercial crops in south India. In Kerala (India), jackfruit tree supports the black pepper vine, which is a climber. Thus, the trunks of jackfruit trees of Kerala are usually covered with the dark green leaves of the pepper vine cultivated by the farmers. The most serious pre-harvest physiological problem in jackfruit cultivation is pre-mature fruit drop. On the other hand chilling injury is a physiological storage disorder.

1. Pre-mature fruit drop

Pre-mature fruit drop in jackfruit is often related to unfavourable environmental conditions, irregular watering, improper nutrition and hormonal imbalance. This causes huge economic losses to the growers every year. For mature trees watering is necessary during dry periods from blooming throughout fruit development, to prevent flower or fruit drop. Growth regulator spray is also found to be helpful in checking hormonal imbalance.

2. Chilling injury

Jackfruits exposed to temperatures below 12°C (54°F) during storage before transfer to higher temperatures exhibit chilling injury symptoms. These symptoms include dark-brown discolouration of the skin, pulp browning, off-flavour development and increased susceptibility to decay. To avoid this disorder, jackfruit can be kept wrapped in polyethylene bags and stored not below 12°C.

❑❑❑

Chapter-23

Jamun

Scientific Name : *Syzygium cumini*

Family : Myrtaceae

Jamun is tall evergreen tree indigenous to India. According to Hindu tradition, Rama subsisted on the fruit in the forest for 14 years during his exile from Ajodhya, because of this, many Hindus regard jamun as a 'Fruit of the Gods'. It is known by different names in different regions of India: these names are java plum, black plum, jambul and Indian blackberry. Though the fruits of jamun are liked by all and sell at a good price, but it has still not grown as an orchard tree and is generally known as an avenue tree or as a wind break. Jamun is a hardy fruit plant which can be grown under adverse soil and climatic conditions. It is one of the hardiest fruit tree and best suited for wastelands. It is drought tolerant and at the same time can tolerate water stagnation as well as marshy wetlands. The bark, leaves, fruit and its seeds i.e. almost all of its parts are used for various purposes that's why jamun tree is considered as another wonderful gift of nature. The wood is used as a timber in building and for manufacturing railway sleepers.

According to the nutritionists the fruit is rich in carbohydrates, minerals (manganese, zinc, iron, calcium, sodium and potassium) and vitamins. It comprises glucose and fructose as principal sugars. The presence of oxalic acid, tannic acid, gallic acid and certain alkaloids makes its taste little bit astringent. The ripe jamun fruit is well recognized as a liver stimulant, digestive, carminative and coolant. Their hypoglycemic property (lowering blood sugar) is well recognized in Ayurveda and Siddha system of medicine in India. The seeds reduce blood sugar levels and glucosuria in diabetic patients. There is no major problem which affects jamun cultivation. The only problem is that it suffers from heavy drop of flowers and fruits at various stages of development.

1. Flower and fruit drop

In Jamun, excessive fruit drop is a serious problem and the fruit drop occurs continuously at various stages of fruit growth and development which results in the considerable reduction in yield and it affects the profits of the growers adversely. This is a serious physiological problem, where flower and fruit drop starts as early as first opening of flowers and continues upto maturity. This may be due to adverse environmental conditions, inadequate soil moisture and hormonal imbalance.

The flower and fruit drop occurs in 3 waves.

A. The first drop takes place during bloom or shortly there after. This is the heaviest drop as about 52 per cent of the flowers drop off after 4 weeks from flowering.

B. About 35-40 days of full bloom, the second wave starts.

C. The third drop takes place after 42-50 days of full bloom and continues till mid July.

Only 12-15 per cent flowers reach maturity as about 65 per cent flower and fruit drop in the first five weeks and since then a maximum of 19-21 per cent flower and fruit drop occurs up to maturity.

Control

a. Proper care and periodical watering must be given during summer months. It will help for bud formation, fruit set as well as it prevents fruit drop.

b. The extent of flower and fruit drop in Jamun may be reduced by two sprays of GA_3 60 ppm, one given at full bloom stage and the other 15 days after initial setting of fruit.

❑❑❑

Chapter-24

Kiwi fruit

Scientific Name : *Actinidia deliciosa*

Family : Antinidiaceae

Kiwifruit or Chinese gooseberry (*Actinidia deliciosa*) is known as 'China's miracle fruit, and 'the horticultural wonder of New Zealand'. From China it spread to New Zealand where it was recognized as a potential fruit and became a popular backyard vine. The name "Kiwifruit" comes from the Kiwi, a brown flightless bird and New Zealand's national symbol. It has a fibrous, dull brown-green skin and bright green or golden flesh with rows of tiny, black, edible seeds. The fruit has a soft texture and a unique flavour and today it is a commercial crop in countries like Italy, China, and New Zealand. It is native to Southern China, where it was known as *Yang Tao* (sunny peach). It is known as "Chinese Gooseberry", because of its visual resemblance to gooseberry.

A ripe kiwifruit is refreshing, delicate, flavoured with pleasing aroma and high in nutritive value. It is mostly eaten as fresh fruit or combined with other fruits in salad and desserts. Fruits can be eaten fresh or processed into many products like jam, squash, juice, preserves and wine. Kiwi fruits are rich in many vitamins (A, C and E), flavonoids and minerals (phosphorus, potassium and calcium). In particular, they contain a high amount of vitamin C (more than oranges), as much potassium as bananas and a good amount of beta-carotene. The kiwifruit skin is edible and contains high amounts of dietary fiber. Kiwifruit is a natural source of lutein and zeaxanthin. Seeds are used for making pastries, fragrant flower is used in producing perfume and roots processed into effective insecticides against aphid and rice borer.

In India, the area under this fruit is negligible being a new exotic introduction. Its commercial cultivation has been extended to the mid hills of Himachal Pradesh and Jammu and Kashmir. Its total production and productivity is very low, yet there is tremendous scope for its cultivation because of wider adaptability, precocity in bearing with high returns, easy to propagate, no serious pests and diseases, earning good price in the market, high nutritive and medicinal value and multiple uses. However, it faces few minor physiological disorder problems such as sun scorch, flats, water stains etc. which are discussed below.

1. Sun scorch or sunscald

Sun scorch or sunscald is a physiological disorder in which fruits are affected due to direct exposure to sun rays and as a result of that the fruit becomes insipid, unfit for human consumption and processing. The affected surface of the fruit develops typical brown, leathery and sunken scars. In summer, high temperature (> 35^{0}C) accompanied by high insulation and low humidity may also cause scorching of leaves.

The damage from sunscald can be minimized by thatching the branches with paddy straw, dry grass or hay as this avoids the direct exposure of fruits to sunlight. To obtain high quality kiwifruit it is necessary that fruits should be free from sunscald and scars.

2. Flats

This disorder is named due to flattened shape of kiwi fruit. This distortation of fruit from its normal shape is due to the development of flattened fruits is quite common in Monty cultivar in India. The reason assigned to this malady is improper pollination. The proportion of flattened fruits in kiwi can be controlled by hand pollination or provision of suitable pollinizers like honeybees in the orchard.

3. Water stain

Water stained kiwi fruits fetch poor price in the market causing economic loss of produce. It is characterized by distinct dark streaks or stains which occurs down the side of kiwi fruit due to deposition of tannins. This accumulation is a result of leaching of tannins from the dead fruit tissues due to rainfall. The malady can be controlled by removal of dead tissue from the plant canopy. The blemishes can be rectified by treatment with weak solution of citric acid.

4. Flesh translucency

Flesh translucency is freezing damage which gets started at the stem end of the fruit and progresses toward the blossom end as the severity increases. As a result of prolonged storage the susceptible fruits become somewhat yellow fleshed and there is lack of "graininess" in such fruits. The early picked kiwifruit when stored at temperatures below 0°C or when subjected to an early frost in the vineyard are more susceptible to this freezing injury of flesh. The damage can be avoided by storing the fruits at safe temperature *i.e.* above 0°C.

5. Internal breakdown

The internal breakdown of kiwi fruit is characterized by a slight discolouration at the blossom end of the fruit. This water soaking progresses around the blossom end and ultimately spreads to the whole fruit resulting in "graininess" below the fruit surface which begins in the area around the blossom end of the fruit.

6. Pericarp granulation

In pericarp granulation the stylar end of the fruit is predominantly affected. The severity of the disorder increases with prolonged storage and after ripening at 20°C. However, storage for optimum time minimizes this defect.

7. Pericarp translucency

Pericarp translucency is characterized by the appearance of translucent patches in the outer pericarp tissue at the stylar end which may extend up the sides of the fruit. This disorder is more severe after prolonged storage, but it can be observed after 12 weeks of storage at 0°C. The symptom development is enhanced by the presence of ethylene in the storage atmosphere. To control this disorder, do not store the fruits for longer times.

❑❑❑

Chapter-25

Litchi

Scientific Name : *Litchi chinensis* Sonn.

Family : Sapindaceae

Litchi (*Litchi chinensis* Sonn.) a native of China, is grown in India since 18th century. It is member of the genus *Litchi* in the soapberry family, Sapindaceae. The name of the fruit is derived from the Chinese word lee chee which means "one who gives the pleasures of life". Litchi is also known as the "fruit of romance" in China. The most important use of litchi is as table fruit in ripe form although it is also used in processed forms such as squash. The fresh fruit has a "delicate, whitish pulp" with a "perfume" flavour. Eating litchi can benefit those suffering from colds, fever and sore throats. Great source of Vitamin C and potassium, it also contains phosphorous, calcium, magnesium and protein. According to a Chinese book, the nature of the flesh of the litchi is warm and it can

Fruit cracking

helps improveing the blood circulation. The Chinese also use litchi for medicinal purposes and for making wine.

India is the second largest producer of litchi in the World after China. Frost during winters and dry heat in summers are limiting factors for its successful cultivation. Harvesting is usually done in May and June. The fruits are harvested in bunches along with a portion of the branch and a few leaves. This helps in improving the keeping quality of the fruits. Litchi is non-climacteric fruit that possesses poor shelf life and therefore needs specific treatment before packing and transportation for long distance market. Litchi being a highly perishable fruit, its marketing should be done as early as possible. For local markets, the fruits should be collected at the ripened stage, while for distant market, the fruits should be harvested when they have started turning reddish. After harvesting, the fruits should be kept in cool place. If exposed to sun even for a few hours the quality deteriorates markedly. The physiological disorders which are commonly observed are fruit cracking, fruit drop and sun burn. These disorders are limiting factors in litchi production and causes considerable losses if are not managed properly.

1. Fruit cracking

The splitting or cracking of fruits is quite common in almost all the litchi growing areas of India, particularly under dry conditions. In severe cases, the crop losses due to this malady may be as high as 50 per cent or even more. Skin cracking of developing fruit is a serious problem in litchi which results in impaired quality fruits unfit for consumption. These fruits fetch poor price in the market giving low returns to the farmers.

Causes

Splitting of fruits has been generally attributed to abrupt changes in atmospheric temperature, relative humidity and soil moisture conditions, following rain or heavy irrigations. The disorder is noted shortly before maturity when prolonged period of drought during which the fruit growth is checked, is followed by excessive moisture supply. Excessive soil moisture aided by fluctuations in temperature and humidity may aggravate fruit splitting. Temperature higher than 38^0 C in combination with relative humidity lower than 60 per cent are very favourable for cracking of litchi fruits. Inadequate moisture during the early period of fruit growth results in hard and inelastic skin susceptible to cracking when subjected to increased internal pressure as a result of rapid aril growth following heavy irrigation after prolonged dry spell. In general cultivars which have relatively thin skin, few tubercles per unit area and rounded to flat in shape are less prone to cracking.

Control

The following measures are generally considered appropriate for adoption to minimize fruit splitting.

a. Regular irrigation in the orchard helps in maintaining growth and expression in the fruit.

b. Frequent irrigation during the critical period of aril growth, spraying of zinc sulphate (1.5%) at weekly intervals, starting from pea stage of fruit growth to harvest reduces the incidence of fruit cracking.

c. In the absence of rains during summer months, water spray may prove useful in keeping the ambient atmosphere of the fruit humid, as moisture has a good local effect on the fruit against splitting.

d. The litchi plants should be trained to keep low headed. Such trained plants having dense foliage can withstand more hot and desiccating winds as compared to tall trees reducing the fruit cracking.

e. Windbreak with one or two rows of tall trees e.g. Safeda alternating with small sized trees having dense foliage (mulberry, jamun, etc) may be planted around the orchard. This helps to create humid conditions in the orchard, which are favourable for the developing fruits and thus splitting is avoided to a great extent.

f. Early varieties like Dehradun and Saharanpur split more than mid season or late varieties. Hence, the varieties which are less prone to splitting should be planted.

g. The litchi orchard should be planted by a hillock or near the large bodies of water or in between tall and dense trees like mangoes.

h. The plants can be guarded immensely against high temperature by sowing Jantar or Arhar along the tree rows or in the periphery of the plant basin. These helps to minimize splitting of fruits, besides protecting the fruit trees from the vagaries of hot weather.

i. Apply some mulch to tree basins during hot summer months to conserve soil moisture.

j. Spray of NAA and 2,4,5-T @ 35-100 ppm has been found effective in checking fruit splitting and increasing fruit size.

k. Spray of Borax (0.8 %) during the fruit growth period is useful to control fruit cracking.

2. Fruit drop

Fruit drop is considered one of the major bottlenecks in the expansion of litchi cultivation in our country. For the optimum crop in litchi, only 4-5 per cent flowers are required. The initial fruit set in litchi is very high but a very small proportion finally matures. The premature fruit drop commences soon after fruit set and continues till fruit maturity, with most fruit abscising in the first 2-4 weeks.

Causes

High incidence of fruit abscission in litchi is mainly stated as physiological rather than a genetic problem. Following are the factors which are resposible for fruit drop in litchi.

i. Competition among fruits for water and nutrients.

ii. Strong desiccating winds.

iii. Failure of fertilization.

iv. Embryo abortion.

v. Internal nutrition.

vi. Hormonal imbalance.

vii. External factors like high temperature and low humidity.

Control

a. Grow varieties like Seedless Late and Calcuttia which are less prone to fruit drop than Dehradun and Muzzafarpur which are more prone to fruit drop.

b. Irrigation to bearing litchi trees twice a week from April onwards and good wind-break minimize fruit drop to a great extent.

c. Foliar spray with zinc sulphate at 0.5, 1.0 and 1.5 per cent considerably increase the zinc content of the leaves and effectively reduce fruit drop.

d. Similarly, spray of NAA and 2, 4-D @ 15 ppm in combination with 1 per cent zinc sulphate results in more fruit retention till harvest.

3. Sunburn

It is a major issue in litchi production in India which poses a threat to growers almost every year. The time of fruit colour change (about 1 month before harvest) is the most vulnerable stage for sunburn. "Sunburn" means a condition on the surface of the skin of a litchi giving it a yellow, brown or black colour and which is caused by excessive exposure to the sun. Sunburn of the fruits can be minimized with proper water management and protecting the fruits with shade nets atleast 30 days before harvesting.

❑❑❑

Chapter-26

Loquat

Scientific Name : *Eriobotrya japonica*

Family : Rosaceae

The Loquat is a fruit of south eastern Chinese origin and it is comparable with its distant relative, the apple, in many aspects such as high sugar, acid and pectin contents. It is mostly eaten as a fresh fruit and it mixes well with other fruits in fresh fruit salads. Pies or tarts are made from firm, slightly immature fruits. A major part of total produce is used for fresh consumption, however, it can also be used for making value added products like jelly, jam, preserves, juice and squash. Loquat syrup is used in Chinese medicine for soothing the throat like a cough drop. The flowers are regarded as having expectorant properties. It is good source of vitamin A, potassium, fiber, carbohydrates and is low in cholesterol.

Loquat was introduced in India under the name of 'Japanese Medlar' and 'Japan plum'. Its commercial cultivation is mostly confined to Uttar Pradesh (Saharanpur, Dehradun, Muzaffarnagar, Meerut, Farrukabad, Kanpur and Bareilly), Delhi, Punjab (Amritsar, Hoshiarpur and Gurdaspur), Himachal Pradesh (Kangra) and to a small extent in Assam, Maharashtra and hills of south India. Since the tree bloom between November and late January at certain places, the crop may be destroyed by moderate winter frosts. The fruit of loquat is most susceptible to frost injury when it just starts colouration and frost is a limiting factor for its successful cultivation. It is available in the market in India during mid-March to May when there is scarcity of fruits and therefore, it is able to bring a good income. Loquat faces some pre-harvest and post-harvest physiological problems resulting in economic losses to loquat growers which are discussed in this chapter.

1. Creasing

Creasing is a skin disorder which consists of grooves or furrows in an irregular pattern on the skin of loquat fruit. Thin-skinned cultivars are more susceptible to creasing. Creasing is caused by formation of abscission zone between the fruit base and the peduncle. The incidence of skin creasing coincides with spring flushing and with the maturation of spring shoots, the incidence of creasing decreases. Although, spring flushing might be the main cause of creasing before harvest, it may also be related to a complex of pre harvest factors, including potassium nutrition, but the causes are not fully understood. Pre harvest sprays of gibberellic acid have been reported to reduce creasing.

2. Internal browning

Internal flesh browning followed by tissue breakdown is enhanced by higher temperatures and longer durations of storage. Elevated CO_2 concentrations (>10%) may induce internal flesh browning and skin brown spotting. Therefore, storing the fruits at optimum temperature and CO_2 concentrations can prevent internal browning in loquat.

3. Russeting

It is the appearance of corky, roughened, brownish or greyish areas on loquat fruits. It is characterized by a number of small cracks which are often concentric and pre-harvest skin blemishes (russeting) may appear on developing fruit. Its severity depends on cultivar, season, and microclimatic conditions. There are several possible causes for fruit russeting, including cool wet weather, frost damage, spray damage etc. The first symptoms of russet appear 30-42 days after fruit set, during the final stages of cell division. The greatest increment in russet however, develops during 54-76 days of fruit development i.e. at the beginning of the cell enlargement phase.

Causes

Following are the main causes assigned to fruit russeting

i. Application of chemicals in excessive quantities under poor conditions as a result of which the cuticle become corky and less elastic. As the fruit growth continues, the cells are not elastic enough and small cracks start to appear with the increase in fruit size.

ii. Large variations in temperature particularly when daytime temperatures are high (accompanied by relatively low humidity) and night temperatures are low. The cuticle cells of fruits are not elastic enough to cope with the changes.

iii. Sensitivity of russeting also varies with different varieties.

Control

Below mentioned control measures are helpful in controlling russeting.

a. Proper chemical concentrations should be used for spraying loquat. Spray only under proper conditions by using the correct equipment.

b. Proper water management practices should be followed to control growth and irrigate at fixed intervals and reduce the frequency of watering when necessary.

c. Fertilizer recommendations should be properly followed to avoid overly succulent plants.

d. Do not remove leaves excessively as leaves create a micro-climate for the fruit which helps prevent russeting.

e. Use varieties that may be more resistant to russeting.

4. Sunburn (purple spot)

Sunburn, "purple spot", is responsible for much fruit loss in hot regions with long summers. During winter if the heat is high, it may also result in sunburned fruits. To produce high quality fruits, the fruit clusters may be bagged so as to protect them from being damaged by sunburn (purple staining of the skin). Chemical sprays which hasten fruit maturity, avoid sunburn. Bagging of the fruit can protect them from this blemish. The best used bags are 2- and 3-ply newspaper bags.

❑❑❑

Chapter-27

Mango

Scientific Name : *Mangifera indica* L.

Family : Anacardiaceae

Black tip of mango

Mango is the national fruit of India which has been cultivated in India from time immemorial. It is indigenous to India and dates back to around 4000 BC. The wild mango is said to have been originated in the foothills of Himalayas in India and Burma. The mango was cultivated as early as 2000 BC. Gradually, the wild mango developed into an exotic, richly flavoured succulent fruit that is enjoyed even today. In the early stages of domestication, fruits were probably very small and fibrous without much flesh. The Mughals and Portuguese selected and grew mango plants for generations. Centuries of development have produced varieties of mangoes free from both fibre and unpleasant flavour. This eventually led to larger fruits with thick flesh that we are familiar with today. The Chinese traveler Huien T'sang, who visited Hindustan between 632 and 645 A.D., was the first person, so far as known, to bring the mango to the notice of the outside world.

Mango is one of the delicious tropical seasonal fruit known as "*King of the Fruits*". It is one of the most popular, nutritionally rich fruit with unique flavour, fragrance, taste and health promoting qualities making it a common ingredient in new functional foods often called "super fruits". Mango fruit is rich in pre-biotic dietary fiber, vitamins, minerals and poly-phenolic flavonoid antioxidant compounds. It is the richest source of vitamin A among fruits and also contain fair amounts of vitamin C. The vitamin C levels in raw mango are higher than that found in a ripe mango. Mangoes also contain traces of vitamin B, E and K. Fresh mango is a very rich source of potassium and copper.

Although, India is the largest producer of mango but in terms of productivity it lags far behind. The low productivity is mainly due to the problems associated with mango cultivation. It suffers from several physiological disorders at all stages of its development, right from the plants in the nursery to the fruits in storage or transit. Some of these disorders take heavy toll on trees and have become limiting factor in mango cultivation in some regions causing great economic losses and these problems are considered as national problems. Major disorders of mango and their control measures are discussed below.

1. Mango malformation

Malformation is a major malady of mango in India causing heavy losses to the mango growers. The incidence of malformation is less in western and southern India, however, almost all the commercial varieties of northern India are highly susceptible to malformation. Maximum incidence has been reported from Punjab, Haryana and western Uttar Pradesh, where more than 50 per cent of the plants are affected by malformation. This is a big threat to the mango industry. Almost all varieties of mango are susceptible to this disorder except Bhadauran, Alib and Illaichi. Mango malformation is of three types-vegetative, floral and mixed. The seedlings and young plants in the nursery are affected by vegetative malformation, whereas, the bearing plants are affected by floral malformation. It is the floral malformation which directly affects the productivity of plants.

A. *Vegetative malformation:* In this type of malformation, a compact leaf bunch is formed at the terminal portion of the shoot or in the axils of leaves and/or on the lower nodes of the seedlings. The buds get swollen and get converted into a thick and short shootlets bearing numerous small scaly leaves. Shootlets arise from the axils of scaly leaves giving abnormal rosette appearance. Many such shoots may arise to form a bunch, hence known as '*bunchy top*'.

B. *Floral malformation:* In floral malformation the panicles (inflorescence) are affected. The floral branches are formed as crowd of cone and compact masses of sterile flowers form a bunch. A profuse development of numerous small, leafy structures occurs in place of flowers, resulting in a *Witch's broom* appearance. The flowers are crowded, enlarged with thick pedicels and calices, enlarged petals and stamens. Fruit set is seldom seen in such flowers. These dry up, remain hanging on the tree and get converted into black masses of dead tissues. This type of malformation has profound influence on the sex expression of mango because malformed panicles mostly produce male flowers and hence no fruit is produced on these panicles.

C. *Mixed malformation:* In mixed type there is development of leafy structures in the panicle making it malformed.

Causes

A large number of factors are considered responsible for malformation which are discussed below.

i. Hormonal imbalance: It has been found that mango malformation is due to imbalance between growth promoters and inhibitors in plants. There is auxin depletion and formation of malformin like substances.

ii. Cultural practices: Poor cultural practices such as unploughed land, improper tilth, etc. lead to this disorder.

iii. Nutritional imbalance: It is due to use of imbalanced fertilizer mixture such as N,P,K and non-availability of the essential nutrients in soil.

iv. Environmental factors: Factors such as unsuitable condition of humidity, temperature, rainfall, etc are all responsible for this malady.

v. Fungal attack: The fungus *Fusaruim moniliformae* var. subglutinans has been isolated from malformed panicles. Fungus may not show any symptoms of disease until there is any injury on top of plant. It also need spreading agent for causing infection.

vi. Attack of mites: It is also considered that attack of mites is another cause of malformation. The injury to the plants provides a mean for entry of pathogen into the host plant.

Control

Following control measures are helpful to reduce the incidence of mango malformation.

a. If the disease is to be controlled chemically then it may be effective only when combination of fungicide and insecticides are used to kill the fungus and the mites.

b. All the infected shoots should be pruned off. Then a mixture of fungicide 0.1% captan, miticide akar 33.8 (0.1%) should be thoroughly sprayed and repeated after every 10-12 days.

c. Balanced fertilizer mixture containing high proportion of N in NPK (9:3:3) reduces the incidence of this disorder.

d. One spray of 200 ppm NAA in October (at the time of bud differentiation stage) is effective in reducing this malady.

e. A single deblossoming treatment in January, at bud burst stage results in increased yield. Early deblossoming combined with NAA spray may reduce the extent of malformation considerably.

f. Use of anti-malformins like glutathione and ascorbic acid are equally useful.

g. Prune off the affected panicles and shoots soon after they appear and bury them in the soil or burn them immediately.

2. Fruit drop

Fruit drop is another serious problem in mango which causes great losses to the mango growers. The magnitude of the incidence can be realized from the fact that hardly 0.1 per cent flowers after pollination and flower set reaches up to maturity. A tree producing several thousand panicles yields only a few hundred fruits. Maximum fruit drop occurs in last week of April or first week of May. The phenomenon of fruit drop in north India has been divided into three distinct phases or waves *i.e.* pin head drop, post setting drop and May drop. Most of the flowers fall down after full bloom or at later stage of development. Drop of young, small fruitlets soon after pollination is sometimes considered useful but the drop of fully mature fruits just before harvest causes considerable economic loss to the growers. Some varieties of mango for *e.g.* Langra show the tendency of fruit drop before it ripe.

Causes

Several factors are responsible for fruit drop. The main causes are as follow:

i. Formation of abscission layer at the point of attachment of fruit with the twig.
ii. Sometimes low carbohydrate content leads to the leaf, flower and fruit abscission.
iii. It may be due to embryo abortion.
iv. Degeneration of ovule may be another cause.
v. Large number of developing fruits is also responsible.
vi. Poor soil also creates conditions for fruit drop.
vii. Inadequate irrigation water leads to fruit drop.
viii. Attack of insect, pest and diseases.
ix. Depletion of nutrients is also a factor.
x. Hormonal imbalance paticularly auxins, etc. are responsible for fruit drop.
xi. Pollination from inferior variety is another cause.

Control

This wasteful malady can be controlled by encouraging following practices.

a. Pollination from amenable variety by planting such plant in orchard.
b. Irrigation is beneficial when the fruiting has taken place.
c. Proper application of nitrogen as it retards and reduces abscission.
d. Application of proper additives to plants.
e. Two sprays of NAA @ 10-20 ppm during April to May is useful.
f. Fruit drop can be controlled to some extent by spray of 20 ppm of 2,4-D in last week of April or in first week of May in Langra and Dusehari cultivars.

3. Black tip

Black tip is another physiological disorder of mango, the causes of which has been attributed to coal fumes and gases of brick kiln by the investigators. The association of black tip with the fumes of brick kilns was first reported by Woodhouse in Bihar in 1909. It is a serious problem in all the states where brick kilns are prominent. In north Indian conditions, the damage due to this malady is alarming, especially near towns and industrial regions. The fumes of brick kiln mainly contain gases like SO_2, CO, CO_2 and ethylene and it was found that from these SO_2 causes the maximum damage.

The different varieties growing in the proximity of the brick kiln show different degrees of damage. Dushehari has been found highly susceptible and Lucknow Safeda, the least. In addition, the velocity of wind and tree vigour also influences the intensity of this malady. The disorder is characterized by depressed spot of yellowish tissue at distant end of fruit which increases in size becoming brown and finally black. The growth stops and after pre-mature ripening such fruits become soft which never reaches maturity and drop down earlier.

Control

i. Shifting of site of brick kiln 2 km on east and west and 1 km on north and south of orchard.

ii. Operation of brick kiln should be avoided from February to 4th week of May.

iii. Telescope chimneys should be 40-50 feet high if at all the operation of brick kiln is necessary during the prohibited period.

iv. Spraying of 0.6 per cent Borax three times *i.e.* before flowering, during flowering and after fruit setting is necessary.

v. Caustic soda 0.8 per cent sprayed twice between last week of March and 3rd week of April minimize the losses to a great extent.

vi. Bordeaux mixture 2:2:250 or 1.5 kg of copper oxychloride per 500 lts of water should be used.

4. Spongy tissue

Spongy tissue disorder was first noticed in Alphonso mangoes by Cheema and Dani during 1932. It is most serious mango disorder in Gujrat and surrounding areas. In Alphonso it is observed to an extent of 35-50 per cent depending upon location, season, age of tree, time of picking, fruit weight, fruit maturity, soil type and environmental conditions. This disorder is also found in Jamadar, Olor, Vellaikolumban, Swarnarekha and Fernandin varieties. This defect is higher in coastal region and at base of hilly areas whereas, its incidence is very less on slopes and in plains. Due to this disorder, the export of delicious Alphonso mango from India has declined considerably.

This malady has become more complicated since the affected fruit presents healthy external appearance. Its symptoms are not apparent at the time of fruit development, picking or when ripe. Affected tissue is visible only when the ripe fruit is cut. A non-edible, sour, yellowish and sponge like patch with bad odour with or without air pocket develops in the mesocarp of the fruit during ripening. This disorder adversely affect the fruit quality making it unfit for human consumption. The symptoms at harvest are localized predominantly at the distal end of the fruit, whereas post-harvest exposure of fruits to the sun around mid-day causes maximum occurrence of spongy tissue in the middle part of the fruit. Such fruits become unpalatable as the affected fruits apart from bad odour have high acidity, low ascorbic acid, carotene and sugar content than the healthy fruits.

Causes

Following are the major causes of spongy tissue in mango.

i. Harvesting the fruits at full maturity increases the incidence of this disorder. It is also considered that the percentage breakdown due to this disorder in Alphonso mango fruits increases with increase in fruit weight.

ii. This disorder appeared to be caused by convective heat rising from the soil rather than by nutrient imbalance or pathogens. Thus, spongy tissue in fruits seems to be a physiological disorder in which the fruit pulp remains unripe because of unhydrolysed starch due to histological and biochemical disturbance caused by heat in the pulp of a mature fruit at the pre-and post-harvest stage.

iii. Close relationship has been observed between the fruit transpiration and spongy tissue. It is also suggested that the lower fruit transpiration rates in cv. Alphonso are a varietal specific trait which results in slower movement of water and minerals to the fruits from soil leading to the development of spongy tissue.

Control

Following practices help to reduce the incidence of spongy tissue in mango.

a. Grow spongy tissue resistant mango hybrids like Ratna, Arka Aruna and Arka Punit which have Alphonso like characters and do not suffer from this malady.

b. Harvesting mangoes at ¾ maturity rather than at full maturity reduces this malady.

c. Orchard management practices like sod culture and mulching are useful in reducing this incidence.

5. Internal necrosis

In internal fruit necrosis browning of pulp in the developing mango fruits are noted. The affected fruits drop down before reaching maturity. The first symptom is the development of dark green colour on the apical part of the fruit and formation of isolated brown areas of indefinite outline in seed and mesocarp of the rapidly growing fruits. The browning of the tissue later on extend to the epicarp and brown black gummy substances exudates from the fruit surface below the green tip. Subsequently, the affected mesocarpic tissues collapse and cavities surrounded by corky tissues develop. The mesocarp splits longitudinally and dark brown portion becomes visible. The disorder gets to halt when fruits attain maturity. Such fruits are easily detachable from the stalk and drop down before ripening. The difference of black tip from internal necrosis is that, in black tip disorder fruits are firmly attached to the tree and are not easily detachable.

In this disorder, browning of developing stone is also very common. It may affect all the cultivars of mango like Dashehari, Chausa, Bombay green, Lucknow Safeda, but Langra is free from this disorder. It is mainly ascribed to boron deficiency in the soil. The foliar (0.5%) as well as soil (500g/tree) applications of Borax decrease the percentage of necrotic fruits. The Borax spray is found to be most effective at the marble stage of fruit development.

6. Soft nose

This disorder was first reported in Florida in Indian origin variety known as Mulgoa. Later it was also observed in varieties originating from Mulgoa variety such as Kent, Haden, Sensation, Tommy Atkins, etc. It is also known as *"insidious fruit rot* or *yeasty fruit rot"*. The typical symptoms are breakdown of the flesh towards the apex of the fruit before ripening. The fruits which are retained on tree for longer time show symptoms of soft nose and such fruits lack firmness. Initially, there is yellowing of green peel of the fruit in the area between apex and the stigma point. On cutting the pulp on the ventral side towards the apex appears to be over ripe, while that around the shoulders and on the dorsal side it is unripe. A mass of yellowish to brown tissue surrounds the over ripe pulp which is bitter in taste.

The incidence of soft nose varies widely with localities and fruits harvested at semi-ripe stage show lesser incidence of soft nose. Also, the incidence is found to be increased by alleviation of nitrogen and calcium level in the trees. Therefore, it can be kept under control by optimization of nitrogen and calcium in soil. Only those varieties should be grown which mature early.

7. Alternate bearing

It is one of the most burning problems, and now is found in many fruit crops and particularly it renders mango cultivation less remunerative to growers. This term is synonymous with 'biennial bearing'. The terms 'periodicity of cropping' and 'irregular bearing' are sometimes erroneously used to describe the phenomenon. In alternate bearing there is production of a heavy crop in one year and no crop or very little crop in the next year. Alternation between good and bad bearing year is not always regular because after a good crop year, two or more poor crop years may follow and conversely after a bad year two or more good crop years may occur.

Causes

Various causes found associated with alternate bearing have been discussed below.

i. *Climatological factors:* Adverse climatic conditions like rain, high humidity and low temperature sometimes convert an on year into an "off" year directly or by promoting the incidence of diseases like powdery mildew and anthracnose. Even frequent frosts adversely affect the fruit set.

ii. *Age and size of shoots:* Vast majority of non-flowering shoots are probably lacking in some vital substance or substances necessary for flower bud formation. The tree therefore takes a year to recoup this loss, thus causing biennial bearing habit. It has also been reported by various workers that in some cultivars it would not differentiate flower buds unless the shoots are of certain length and girth and with a number of leaves of a particular size. This necessarily means that a certain amount of physiological maturity is a pre-requisite to flowering in the case of cultivars which are prone to this phenomenon.

iii. *Carbon/nitrogen ratio:* It has been found by various workers that irregular bearing in mango is caused by nutritional deficiency, especially that of nitrogen. It was further found that the proportionate increase in nitrogen led to vegetative growth whereas, its proportionate decrease favours flowering. It is quite evident that higher starch reserve, total carbohydrates and C/N ratio favour flower bud formation and accumulation of these compounds may create favourable conditions for the synthesis and action of the substances responsible for flowering.

iv. *Hormonal imbalance:* It has been found that alternate bearing is due to imbalance between growth promoters and inhibitors in mango plant.

v. *Varieties:* Biennial bearing is a characteristic of a variety and some workers emphasized that this is due to the genomic constitution of the particular variety. Most of the commercial varieties of north India e.g. Dashehari, Langra and Chausa are prone to biennial bearing but varieties of south India like Totapari Red Small, Neelam and Bangalore are regular bearers.

Control

The horticulturists since decades have worked to control this malady but so far no concrete control measures have been put forward. However, some control measures have minimized serious problem of alternate bearing to some extent. Biennial bearing problem has been a major bottleneck in the expansion of the mango industry. The growers have to split the profit of an "on" year into two, in order to take care of the "off" year. Also because of glut in the "on" year, the price offered to the orchardist is less remunerative. During an "off" year or a year of low production, the consumer too has to pay an exorbitant price for the fruit. Obviously, this problem has, therefore, posed a major challenge to the horticultural scientists during the last few decades.

Consequently, a number of remedial measures have been proposed from time to time to overcome this limitation in mango production. These are as follows:

a. *Proper upkeep and maintenance of orchards:* A proper cultural schedule is of paramount importance for maintaining fruit trees in healthy and disease free condition and thereby, obtaining better plant performance. Proper maintenance of trees may help in reducing erratic or irregular bearing.

b. *Deblossoming:* The practice has been recommended by some earlier workers with a view to obtain some crop every year by reducing the crop load in the "on" year and it put forth panicles in the following year which would otherwise be an "off" year.

c. *Smudging and chemical regulation:* Induction of flowering in mango through smudging (building up slow fires, emitting smoke) is an age old practice in the Philippines. Some attribute the flowering to heat caused by smudging whereas the others hold CO_2 responsible for this. Recent work to control alternate bearing has shown possibility of using an ethylene releasing compound, ethephon, inducing flowers every year. The concentration of ethephon @ 200 ppm coupled with 0.1 per cent urea has been found effective in inducing regular flowering over the year. Five sprays in all are recommended to be made beginning from middle of September at monthly interval. More recently promising results have been reported on the induction of flowering in mango with paclobutrazol. It can be applied in the form of spray or as soil application. As direct application it can be applied @ 2.5-15 g a.i./ tree and as a spray 500-3000 ppm.

d. *Pruning:* Pruning can be helpful in overcoming the problem of irregular bearing in mango. The pruning was recommended for opening the centre of the trees by topping off or thinning of branches which help in reducing irregular bearing. The practice can be helpful in permitting light to old and neglected orchards and in improving their erratic fruiting.

e. *Growing regular bearing cultivars and hybridization:* Most of the south Indian varieties like Totapari Red Small, Neelam and Bangalore are regular bearers. Scientists performed various experiments on hybridization of north Indian irregular bearing varieties with comparatively regular bearing varieties from south India. The scientists have evolved highly regular hybrids Amarpalli (Dashehari X Neelam) and Mallika (Neelam X Dashehari).

8. Clustering (Jhumka)

The disorder was first observed during 1984 in U.P., India. This malady is characterized by a cluster of fruitlets at the tip of the panicle giving an appearance of bunch tip called Jhumka. These fruitlets are dark green with a deeper curve in the sinus beak region compared with normally developing fruitlets. These fruitlets grow to marble size after which their growth ceases.

One of the main reasons for clustering is the adverse climate during February-March, particularly the low temperature. Most of the fruits are aborted with shriveled embryos. Jhumka is attributed to lack of pollination and fertilization due to unfavourable weather conditions and spraying of insecticides during flowering. The other cause is diversion of photo-assimilates to the vegetative growth from the fruits resulting in their drop.

The incidence of Jhumka can be reduced by spray of 200 to 300 ppm NAA during November and lesser use of pesticides in the mango orchards during full bloom. Even for proper pollination during flowering there must be adequate population of houseflies.

9. Leaf scorch

Scorching of leaves is very common in mango during winter season especially in north India. It can be identified by the development of brick red colour towards the tip and along the margins of old leaves. Ultimately, the tissue collapses into necrotic patches. The foliage gives burnt look due to which the symptoms are so distinct that scorched trees are spotted from a distance. It is caused by chlorine deficiency which results in unavailability of potassium to plants. In severe cases, defoliation occur reducing tree vigour and yield. The disorder occurs due to saline soils and excessive use of muriate of potash.

Using sulphate of potash as source of potassium can check leaf scorch effectively. Further, avoiding brackish water for irrigation can control scorching. In acute problem 4-5 foliar applications of potassium sulphate on fresh growth flushes at fortnightly interval greatly reduces the malady.

10. Taper tip

In this disorder, usually the distal end of the fruit is affected showing intensification of the normal green colour in fruits tapering abruptly which are often curved. The affected fruits are smaller in size and are easily detachable from the stalk. It has been observed in cultivars like Dashehri, Bombay Green, Bombay Yellow and Lucknow Safeda however, Dashehri is most affected by this malady.

11. Tip pulp

The fruits of Lucknow Safeda are most affected by this disorder showing yellowing of tip which later turns to grey colour. The pulp becomes soft and pulpy giving slightly sweet but different taste than healthy fruits. The other parts of fruit remain compact, unripe and hard affecting storability of fruits causing complete loss to the growers.

12. Girdle necrosis

In girdle necrosis, the lower half of the fruit gets disfigured with small etiolated spots and the upper half of the fruit is characterized by brown dotted etiolated area which later on collapse and enlarges to form necrotic lesions. The fruits remain under sized and the lesions extend to form necrotic girdle of tissue around the sinus region of the fruit. In initial stages the tip remains green, however, in advance stages tip also show necrosis. In girdle necrosis, mesocarpic tissue is affected prior to epicarpic tissue as there is formation of cavity underneath the necrotic spots. It occurs at any place of the sinus region, forming an annular ring at the sinus region leaving the fruit tip healthy. In this sense it differs from black tip disorder though in advance cases the difference gets minimized as tip also turns necrotic.

13. Fruit pitting

Fruit pitting disorder is characterized by formation of small sunken pits in equal proportion on the peel of the fruits on all the sides when they have half grown. This gives an ugly look to the fruits affecting the consumer acceptability. The pits increase in size as the fruit reaches maturity. Almost all the varieties are affected but Dashehri is most affected. The malady is result of boron deficiency and poor orchard management. The problem can be checked by application of Borax @ 400-500 g/tree to full bearing trees and practicing proper orchard management.

14. Jelly seed

In this disorder the pulp surrounding the stone disintegrates completely turning to jelly like mass. In severe cases there is development of internal cavity and affected fruits when cut give bad odour. The disorder is detected only when fruits are cut as they look normal externally. However, in severe cases the beak is soft to touch. This disorder is aggravated by over fertilization with nitrogen so to control this problem, the rate of nitrogen application is reduced. Even fruits harvested at mature-green stage are less affected than those allowed to ripen on the tree.

❑❑❑

Chapter-28

Mangosteen

Scientific Name : *Garcinia mangostana* L.

Family : Guttiferae

Mangosteen named after the French priest and explorer Laurentiers Garcin is thought to have originated in Southeast Asia. It was said to be Queen Victoria's favourite fruit. The correlation of Queen Victoria with the mangosteen is probably related to the fact that numerous people have referred to it as the "Queen of Fruits". It is one of the most praised tropical fruits and certainly the most esteemed fruit. Most people enjoy mangosteen and the fruit now categorized as a "super fruit" has a ready market in western countries where it is considered a tropical delicacy. Mangosteen is primarily consumed as a fresh fruit. The volume of production is increasing and the fruit is now being processed into value-added products such as jam, candy and wine. Unique for its appearance and flavour mangosteen is often revered as the exotic tropical fruit, quite popular for its juicy, delicious arils all over the Asian countries. This fruit is very low in calories, contains no saturated fats or cholesterol, but rich in dietary fiber. Fresh fruit is a very good source of Vitamin C, B-complex vitamins such as thiamin, niacin and folic acid. Mangosteen also contains a very good amount of minerals like potassium, manganese and magnesium. The rind is rich in pectin and mangosteen twigs are used as chew sticks in Ghana.

The fruit has a good postharvest life which is beneficial for export markets. The main problems in mangosteen cultivation are slow growth and long juvenile period, low productivity, short shelf life of fruit and low fruit quality due to transparent flesh disorder (TFD) and gamboge disorder (GD). The incidence of transparent flesh disorder (TFD) and gamboge disorder (GD) are both major problems of mangosteen production in the humid tropics. Consequently, these disorders downgrade the fruit quality which affects its marketability and price both in the local and international markets as well.

1. Gamboge

It is a serious disorder of mangosteen characterized by discolouration of the pericarp and the fruit pulp by yellowish resin which turns reddish brown and gets harden later. Affected fruit exude yellow latex which is responsible for the discolouration and change in taste. As a result of broken latex secretory ducts caused by excessive rain, winds, careless fruit handling which induce physical damage to the fruit pericarp and also due to some pest infestation, the bitter yellow resin permeates into the aril. Heavy and continuous rains at the time of fruit ripening promote gamboge. During the rapid fruit growth period the latex secretory ducts are usually broken at 5-15 weeks after anthesis. The seed and aril growth rates are reported to be faster than the pericarp causing a building up of internal pressure to the endocarp. The resultant pressure causes the secretory ducts to break especially when the epithelial cells of the ducts are weak.

However, this physiological problem is rectifiable by application of adequate calcium nutrient to mangosteen trees. The applications of calcium chloride spray effectively reduce yellow spots either on the pericarp or the aril of mangosteen. It is also suggested that the gamboge incidence could be avoided by maintaining mild soil water deficit in pre-harvest periods.

2. Flesh translucency/Transparent flesh discorder (TFD)

The incidence of transparent flesh disorder (TFD) is caused by excess water during the pre-harvest stage of fruit development, where water-uptake is driven by the difference in osmotic potential between that of rain water and that of the affected fruits. It is found that the threshold of TFD incidence of mangosteen fruits subjected to excess water is around 9 weeks after bloom. Flesh translucency of mangosteen fruit shows as a water soaking of the flesh. Under severe conditions of excess water, fruit cracking is also found in fruits with translucent flesh disorder. The incidence of TFD in mangosteen can be avoided by the management of soil moisture during pre-harvest times.

❑❑❑

Chapter-29

Olive

Scientific Name : *Olea europaea*

Family : Oleaceae

The olive is native to the Mediterranean region, tropical and central Asia and various parts of Africa. Historically, olive tree was symbol of *'Peace and Goodwill'*. Olive oil has long been considered sacred; it was used to anoint kings and athletes in ancient Greece. It was burnt in the sacred lamps of temples as well as being the "eternal flame" of the original Olympic Games. Olive oil produces many health benefits when used in cooking or when poured over salads. The use of olive oil can improve digestion and can benefit heart metabolism through its low content of cholesterol. Experts claim that olive oil consumption can cause a person to grow shiny hair, prevent dandruff, wrinkles, dry skin and acne, strengthen nails, stop muscle aching, lower blood pressure and cancel out the effects of alcohol. The nutritional value of olives also includes the high vitamin E content. Olives also have high amounts of polyphenols and flavonoids which also act as antioxidants. Apart from this olives are good source of iron, copper, and dietary fiber.

Olive trees are long-lived with a life expectancy of 500 years, survive droughts and strong winds. Olive cultivation is gaining popularity day by day. The olive tree survives even extended dry periods due to its small leaves, with their protective cuticle and slow transpiration. Because the olive has fewer natural enemies than other crops and the oil in olives retains the odour of chemical treatments it is one of the least sprayed crops. Few physiological disorders have to be checked upon to make its cultivation remunerative. These include fruit drop, chilling injury and fruit spotting.

1. Fruit drop

Like many other crops, olive is also prone to fruit drop. It is a complex phenomenon which occurs due to large number of factors like variety, tree aspect, environmental conditions, nutritional condition of the plant, number of developing fruits and hormonal imbalance particularly auxin. However, extreme heat during critical growing periods can lead to excessive fruit drop in olives. Lack of soil moisture during these periods results in abscission of fruits. Heat stress results in wilting showing a sign of malfunctioning at optimal levels with an indication of reduced photosynthesis. Due to this reason the fruits get starved for nutrients and thus, do not reach their full maturity. In extreme cases, the fruits drop from the tree. Fruit drop can be controlled by maintaining optimum soil moisture.

2. Chilling injury

Chilling injury or internal browning is a major storage disorder of olive caused due to exposure of fresh olives to low temperature. The incidence and severity of disorder depends on storage temperature and duration as well as cultivar. It can be a major cause of deterioration of fresh olives stored at low temperatures. The injury occurs if olives are stored longer than 2 weeks at 0°C or 5 weeks at 2°C before processing. As the time progresses internal browning begins in the flesh around the pit and radiates outward toward the skin.

However, it can be avoided by exposure of fresh olives to temperatures below 5°C (41°F). The ideal storage conditions are 5 to 7.5°C (41 to 45°F) and 90-95 per cent relative humidity.

3. Fruit spotting and marking

Fruit spotting and marking may possibly be due to spray damage and bruising during harvest. It degrades the quality of fresh olives as they are used for topping in many Italian dishes. So, it influences consumer acceptability. The fruits should be sprayed and harvested carefully to avoid such problems.

❑❑❑

Papaya

Scientific Name : *Carica papaya*

Family : Caricaceae

The papaya, papaw or pawpaw is the fruit of the plant *Carica papaya*, commonly known as 'melon tree' in the genus *Carica*. It is native to the tropics of the America. This highly loved tropical fruit was reputably called "The Fruit of the Angels" by Christopher Columbus. Papaya is considered powerhouse of nutrients as it is rich in anti-oxidants, vitamins *viz.* A, B, C, minerals like potassium and magnesium and fiber. Together, these nutrients needed for function of a healthy immune system, promote the health of the cardiovascular system and also provide protection against colon cancer. In addition, papaya contains the digestive enzyme, papain, which helps in digestion and is widely used in industry for various purposes.

Papayas are ready to harvest when most of the skin is yellow-green. Papaya is a thermo-sensitive plant. Temperature is one of the most important climatic factors which determine the success of papaya cultivation. It is very much sensitive to frost and strong winds. Watering is the most critical aspect in raising papayas and it cannot bear water stagnation. The environmental factors profoundly affect the physiological disorders in papaya and an understanding of their interaction with physiological processes is extremely important for economically sustainable production in the nursery or field plantation. Main physiological disorders affecting papaya are pulp gelification and skin freckles.

1. Skin freckles

The cause and factors influencing skin freckles disorder are unknown. Freckle like blemishes occur on ripe papaya fruits but young fruits (less than 40 days old) are free from freckles. Freckles are initially noticed as small pinpoint spots on fruits that are half developed. As the fruit matures the spots slowly increase in size during the last phase of fruit growth as the fruit approaches maturity. The spots are brown in colour, have a reticulated pattern and may have a water-soaked margin. The centers of large spots may attain a greyish colour. More freckles are seen on the exposed side of the fruit away from the stem. They occur throughout the year but appear to be more prevalent during seasons or periods when sunny days prevail.

Freckles are superficial and do not affect the flesh and are therefore, primarily termed as a cosmetic disorder. It is proposed that the high calcium content associated with high temperature and large diurnal temperature fluctuations contribute to greater cell wall rigidity. This facilitates increased turgor pressure resulting in the rupture of the latex vessels and latex leaking which results in skin freckles. Wrapping young fruits in white paper bags significantly reduce freckle incidence.

2. Pulp gelification

Pulp gelification is another physiological disorder of papaya which deteriorates the quality of papaya fruits. It occurs due to magnesium and calcium deficiencies in the pulp. However, controlling such deficiencies can check gelification of the pulp.

❑❑❑

Chapter-31

Passion Fruit

Scientific Name : *Passiflora edulis*

Family : Passifloraceae

The purple passion fruit (*Passiflora edulis*) is a native of the rainforest margins in the Amazon region of Brazil. Passion fruit acquired its name from Spanish missionaries who thought that parts of the plant's flower resembled different religious symbols such as the crown of thorns. Passion fruit grows on a vine in tropical regions. It has a tough, smooth, waxy rind that can be either yellow or purple. The flavour is like a cross between melon and guava. Passion fruits get sweeter as they shrivel. The fruit will quickly turn from green to deep purple (or yellow) when ripe and then fall to the ground within a few days. They can either be picked when they change colour or gathered from the ground each day.

The passion fruit is known for its various health benefits due to its rich constituents like vitamins (A and C) and minerals (iron and potassium). Their seeds are edible and are an excellent source of fiber. It has a pleasant taste and appearance and like its name, it can add passion in one's life due to its various health benefits which aid in the overall well-being of a person. The fruit is valued for its pronounced flavour and aroma which helps not only in producing a high quality squash but also in flavouring several other products. The juice of passion fruit with an excellent flavour is quite delicious, nutritious and liked for its blending quality. To enhance the flavour of the final produce, passion fruit juice is often mixed with juices of pineapple, mango, ginger, *etc.* The juice is extensively used in confectionery and preparation of cakes, pies and ice cream.

In India it is found growing wild in many parts of Western Ghats such as Nilgiris, Wynad, Kodaikanal, Shevroys, Coorg and Malabar as well as Himachal Pradesh and North Eastern States like Manipur, Nagaland and Mizoram. Passion fruit is generally not consumed as a table fruit due to numerous small, hard, dark brown seeds in fruit and its commercial value lies in its processing in preparation of juice, concentrate, squash, icecream, confectionery etc. or in blending its juice with other fruit juices to enhance the flavour. Commercial processing of yellow passion fruit yields 36 per cent juice, 51 per cent rind and 11 per cent seeds. There is very good demand of juice/concentrate in foreign markets. The passion fruit is confronted with some disorders which occurs due to high or low temperature resulting in fruit deterioration.

1. Chilling injury

The symptoms include surface and internal discoloration, pitting, water-soaked lesions, uneven ripening or failure to ripen, off-flavor development and increased decay incidence. Chilling injury occur on passion fruits kept at 5°C (41°F) or lower. The injury can be controlled by storing the fruits above 5°C.

2. Heat injury

Heat injury results from exposure of fruits to direct sunlight or to excessively high temperatures. In passion fruit, it is characterized with shrivelling, bleaching, surface burning or scalding, uneven ripening, excessive softening and desiccation. The affected skin shows brown discolouration. This can be avoided by shading the vines.

❑❑❑

Chapter-32

Peach

Scientific Name : *Prunus persica*

Family : Rosaceae

The word *persica,* along with the "peach" itself and its cognates in many European languages, derives from an early European belief that peaches were native to Persia. Alexander the Great and his armies found the peaches in Persia and brought them to Greece. Most of the peaches grown in commercial orchards today are fruits that are harvested while too firm with a seed that clings to the pulp called a "clingstone" peach. The best flavoured peaches ripen soft and the seed easily separates from the edible portion, and these are called "freestone" peaches.

Peaches are cultivated throughout warm temperate and subtropical regions of the world. In the peach fruit, the stone is covered with a fleshy substance that is juicy, melting and of fine flavour when matured and mellowed. It is rich in proteins, sugar and vitamin A, Vitamin B_1, B_2 and niacin. Peaches also contain the minerals like calcium, phosphorus, iron and potassium.

Introduction of the cultivated peaches into India probably took place in the later half of the 19th century. Today, it is being grown in the mid-hill zone of the Himalayas extending from Jammu and Kashmir to

Khasi hills. The low chilling peaches are grown in sub-mountaneous region and Punjab, Haryana, Delhi and the western Uttar Pradesh. To get premium price and reduce the losses during packaging and transporting, peaches should be harvested at optimum stage of maturity. There are many physiological problems of peaches which should be controlled to get good economic returns. These are discussed below.

1. Sunscald

This is the most destructive physiological disorder of peach where often scaling and peeling off of the bark occurs. Sunscald causes severe damage to the exposed trunk and main scaffold branches. Shading of branches considerably reduces the incidence. Painting of exposed surface with lime paste and shading by wrapping straw or hay around the trunk and thicker branches is quite effective in mitigating the problem.

2. Splitting of fruits and gumming

Splitting generally occurs at dorsal and ventral sides mostly at the time of pit hardening stage. Sometimes gum exudes from the fruit making it unfit for consumption. Splitting and gumming is accentuated during heavy rains after a long dry spell. The gum may fill the entire pit cavity surrounding the kernel. Seed abortion is frequently associated with gumming. The development of split pit is favoured by the conditions which result in development of larger fruits. This causes great loss to the growers as the affected fruits become unmarketable. To avoid this malady, irrigate the orchard frequently to avoid long dry spell.

3. Freezing injury

Freezing injury is characterized by glassy, water soaked or translucent areas in the flesh and with the passage of time these injured areas dry up leaving open "gas pockets" in the flesh. The freeze injured tissue begins to brown as a result of enzymatic oxidation of phenols released by the injured tissue. When freezing occurs at the fruit surface, the glossy or browning symptoms may be visible without cutting. A fruit freezes because of prolonged exposure to a temperature just below its freezing point. Therefore, to control freezing injury maintain temperatures just above freezing

4. Internal breakdown

It is characterized by flesh internal browning, mealiness, bleeding, failure to ripen and flavour loss. These symptoms develop during ripening after a cold storage period, thus, are usually detected by consumers. Fruit stored within the 2.2-7.6°C (36-46°F) temperature range are more susceptible to this disorder. To avoid this physiological disorder the storage of fruit at safe temperature limit is recommended.

5. Inking or black staining

It is a physiological problem affecting only the skin of peaches which is characterized by black or brown spots or stripes. These symptoms appear generally 24-48 hours after harvest.

Cause

Inking occurs as a result of abrasion damage along contamination with heavy metals such as iron, copper and aluminum. This occurs usually during the harvesting and hauling operations, although it may occur in other steps during post-harvest handling.

Control

Inking or black staining can be controlled by following the below mentioned practices.

a. Reduce fruit abrasion damage,
b. Treat fruit gently.
c. Avoid long hauling.
d. Keep picking containers dirt free.
e. Reduce contamination of fruit.
f. Harvesting equipment should be kept clean.
g. Avoid dust contamination on fruits.
h. Check water quality for contamination with metals (Fe, Cu & Al).
i. Do not spray foliar nutrients containing Fe, Cu or Al during fruit maturation.
j. Stop pre-harvest applications of the fungicides and foliar nutrients within 15 days before harvest.

❑❑❑

Chapter-33

Pear

Scientific Name : *Pyrus pyrifolia* (Chinese pear) and *Pyrus communis* (European pear)

Family : Rosaceae

Chinese pear (*Pyrus pyrifolia*) is a pear tree species native to China, Japan and Korea. The tree's edible fruit is known by many names, including Asian pear, nashi or nashi pear, Korean pear, Japanese pear, Taiwan pear, sand pear or apple pear. In South Asia, the fruit is known as *Nashpati*. The tree is a popular symbol of spring in East Asia and is a common sight in gardens and the countryside. The European pear (*Pyrus communis)* is a species of pear native to Central and Eastern Europe and Southwest Asia. The European pear is one of the most important fruit of temperate regions.

The light colour flesh of pears is juicy, sweet and usually mild. Its texture is soft and buttery but some varieties have grainy flesh. Pears are usually thought as bell-shaped but some varieties are shaped almost like a rounded apple. The skin varies in colour viz. yellow, green, brown, red or a combination of any of these colours. Pears are excellent source of water-soluble fiber. They contain vitamins A, B_1, B_2, C, E, folic acid and niacin. It is also rich in copper, phosphorus and potassium, with lesser amounts of calcium, chlorine, iron, magnesium, sodium and sulphur. Pears have anti-oxidant and anti-carcinogen glutathione which help prevent high blood pressure and stroke.

In higher altitude conditions high chilling requirement varieties like Bartlett are mainly grown. In more recent years, red-colour strains of pear like Max Red Bartlett, Red Bartlett and Starking are replacing yellow coloured cultivars. In warmer sub-mountainous areas of H.P. and sub-tropical Punjab, oriental pear cultivars like Baggugosha, Kieffer and Patharnakh are cultivated commercially both for table and processing purpose. Pear is very delicate fruit and it needs lot of care. Careless harvesting practices, whereby fruit is allowed to become overmature before harvest or is delayed in the field or orchard before it is placed in cold storage, and poor storage and shipping operations, which allow temperatures in cold storage or common storage to become higher than those recommended for maximum keeping quality or permit transit delays under unfavourable conditions, all may drastically shorten the marketing life of pears. Produce that has been mishandled is likely to degenerate before the end of the normal marketing season. This bring on physiological deterioration and thus cause unnecessary losses.

1. Brown heart

Premature ripening begins with pink colouration near the blossom end. Consequently, core breakdown *i.e.* brown heart and softening occur in affected fruits which do not ripen properly. This disorder is caused by abnormally cool growing season preceding harvest. Night temperature lower than 7.1°C and day temperature lower than 21°C for a few days are sufficient to cause premature ripening. As soon as the initial symptoms appear, the fruits should be harvested and handled normally.

2. Core breakdown

Core breakdown occurs mostly in European pears viz. Bartlett, Bosc and Comice. The incidence increases in fruits harvested beyond optimum maturity and promoted by delays in cooling after harvest, warmer storage temperatures than recommended and extended storage. It is characterized by brown, soft breakdown of the core and surrounding tissues which may develop in storage or soon after transfer to warm temperatures. In Bosc pear, symptoms may be preceded by vascular browning as the fruit ripens. In Bartlett, the watery, brown tissue separates easily from the healthy tissue. This disorder can be avoided by storing pears beyond their postharvest life. To extend postharvest life, fruit should be rapidly cooled and stored at the lowest non-freezing temperature (generally -2 to -1°C, 28 to 30°F). Controlled atmosphere storage can extend the post-harvest life as much as two-fold. High fruit calcium content has been correlated with reduced susceptibility.

3. Internal browning

Internal browning is a worldwide problem of Asian pears and it is one of the main consumer complaints. The symptoms include the development of brown to dark brown water-soaked areas in the core and even in flesh during storage. There is no visible external indication of internal browning. The main cause of internal browning is that ripening predisposes the fruit to the disorder while it is in cold storage. Internal browning can be avoided by harvesting Chinese pears earlier. The fruit should be picked when most of the pears on the tree are still green, although a few at the top may begin to develop some light-yellow spots. Fruits picked when the skin is completely yellow will develop internal browning within one month after harvest. Prompt cooling of fruit is recommended as delayed cooling increase the incidence of internal browning in pears that are beginning to turn yellow.

4. Senescent scald of European pear

It is common disorder of Bartlett and Bosc pears. It develops on pears which have been stored beyond their post-harvest life. It is characterized by brown to black discolouration of the skin associated with fruit which becomes yellow in storage and loses the capacity to ripen normally. Browning is initially restricted to the skin but progresses rapidly into the flesh, particularly upon transfer of the fruit to warm temperatures. Brown skin becomes weak and is easily sloughed off in later stages of senescence. Taste and odour of the fruit are very disagreeable even before discolouration begins. To control the disorder, avoid storing the fruit beyond its postharvest life. To extend the post-harvest life maximally, harvest fruit at optimum maturity for the length of storage, cool fruit rapidly and thoroughly after harvest and store at the lowest, non-freezing temperature (often -1 to -2°C, 28 to 30°F). Controlled atmosphere storage can extend the post-harvest life of the fruit up to twice that is achievable under air storage. Watch for change in fruit colour from green to yellow. Yellowing is an indication that the fruit's storage life is nearing the end.

5. Watery breakdown of European pear

Watery breakdown can develop in fresh pears, especially the 'Bartlett' variety, during storage and/or ripening following even short periods in cold storage. Watery breakdown can involve any part of the fruit, being most common in the tissue around the blossom end of the fruit. Watery breakdown as the name implies, is a soft, watery deterioration of affected tissue. When the affected tissue is cut or punctured, juice flows out of the fruit. Typically, the involved tissue is in the outer portions of the flesh, but in severe cases can move into the core tissue. The affected tissue is not discoloured during early stages but becomes brown with time. Affected fruits are not useable for either processing or fresh eating and leakage from affected fruit can spoil surrounding healthy fruit.

It is thought that the random pattern of symptom development on individual fruits is related to other stresses that occur during handling. The soft, watery breakdown results from softening of the affected tissue. Fruit that are cooled slowly and/or held at too warm temperature are most severely affected by this disorder. Fruit of any maturity can be affected but high maturity fruit appear to be most susceptible. Cool fruit as soon as possible following harvest and hold them at as low temperature as possible without danger of freezing. To minimize the incidence of watery breakdown fruit should be cooled to near storage temperature within 1 day of harvest and held at 0°C (32°F) or below for up to one month storage or -1°C (30°F) for longer storage.

6. Flesh spot decay (FSD) of Japanese pears

It is characterized by partial browning of spots and development of cavities in Asian pear flesh. It appears along and around the vascular bundles when the symptoms are severe, but there is no external indication of the disorder. Generally, FSD is more pronounced above the equator of the fruit (towards the stem end), but it can also be observed all the way down to the calyx. Cavities are usually dry and surrounded by apparently healthy tissue. This disorder can occur in fruit while still on the tree. It is more obvious, however, after 2-6 weeks of cold storage. The cause of FSD is still unknown, however, climatic factors, such as a fluctuating hot and cool summer or high rainfall right before harvest may enhance the incidence of this disorder.

There is no effective way to control FSD since definite causes have not been identified. The problem is the inability to predict or diagnose

FSD without cutting the fruit. Avoid whenever possible the following conditions that might induce FSD.

a) Low crop load and large fruit.

b) Later picking and advanced maturity.

c) Sunburn and extreme temperature changes during the maturation season.

d) Erratic irrigation or precipitation frequency, amount and timing.

e) Harvesting fruit under warm temperatures and cooling the fruit rapidly.

7. Hard end

This disorder occurs when oriental rootstock is used for commercial cultivars of pear. It is characterized by hardening and blackening of fruit at the blossom end as it approaches maturity. Therefore, the disorder is known as 'hard end'. The malady has been attributed to unfavourable water relationship between the fruits and other plant parts. The problem can be solved by using European pear as rootstocks.

8. Pink calyx or pink end

The development of pink colouration near the blossom end of the fruit is the initial symptom. Later, core breakdown and softening occurs in the affected fruits which do not ripen properly. The symptoms are induced due to the normal growing seasons preceding harvest such as night temperature lower than 7.1°C and day temperature lower than 21°C for few days. This malady appears when premature ripening begins. To avoid the occurrence of disorder, pick and handle the fruits properly without causing any injury and then store and ripen to appropriate quality.

❑❑❑

Pecan nut

Scientific Name : *Carya illinoisis*

Family : Juglandaceae

The English term pecan comes from the Algonquian Indian word paccan or pakan, meaning a nut so hard that it had to be cracked with a stone or a nut requiring a stone to crack. Pecan nut is native to North America and United States produces about eighty per cent of the world's pecans. The Spaniards called it a "type of walnut". Subsequently, the French turned it into "*La Pacaniere*". Eventually, the name pecan nut was created. The nuts of the pecan are edible, with a rich buttery flavour. They can be eaten fresh or used in cooking, particularly in sweet desserts.

One of the most significant facts of pecan nutrition is that these nuts are the best antioxidants. Just a handful of pecans offer vitamins (A, B and E), minerals (calcium, magnesium, potassium and zinc), fiber and more antioxidants than any other nut. Its kernels have high nutritional and calorific value, *i.e.* 11-12 per cent protein, 70 per cent fat and good amount of phosphoric acid. Pecans are also a rich source of oleic acid. The unsaturated fat content in pecan is supposed to be the heart-healthy fat. Sixty per cent of the fat content in pecans is monounsaturated and the leftover thirty per cent is polyunsaturated which means that the saturated fat content in pecan nuts is very low and the unsaturated fats are heart-healthy. So, pecans are called 'heart-healthy' nuts. The wood of pecan nut tree is used in making furniture, in wood flooring, as well as flavouring fuel for smoking meats.

In India, pecan was introduced in 1935 at the Government Progeny-cum-Demonstration Orchard, Palampur, Himachal Pradesh. Pecans are now well-acclimatized in Solan, Mandi and Kullu districts of Himachal Pradesh, hills of Uttar Pradesh, Jammu and Kashmir and Nilgiri hills. Although, there is only limited area under its cultivation but the total pecan nut production is increasing due its high economic returns. Some of the physiological problems that hinder its successful cultivation are discussed below.

1. Rosette

Rosette is characterized by bronzing and crinkling of the leaflets at the initial stage. The leaf finally shows rosette appearance due to extensive crinkling. When the incidence is severe, twig and branch dieback occurs resulting in retarded growth and development of the trees. The affected trees are devoid of nut production. This is a nutritional disorder of pecan nut resulting from zinc deficiency. So, it can be corrected with foliar spray of zinc sulphate (0.5%).

2. Mouse ear

The disorder is characterized by shortening of the mid-vein of the leaflets. The leaflets ultimately become round and wrinkled forming a cap. The leaflets develop into a mouse ear shape and the entire leaf is smaller than the normal size. It is due to manganese deficiency. The disorder can be corrected by maintaining appropriate level of manganese in the soils to avoid its deficiency. Building of high levels of Ca, Mg and Zn in the soil should be prevented.

3. Leaf scorch

Leaf scorch, also known as necrosis is serious disorder of pecan nut. The leaflets show development of necrotic areas on their basal edges. Finally, with the spread of dead areas, tree starts showing signs of defoliation. In summer or early fall, there is poor filling of nuts resulting in a complete crop failure. The main cause of this physiological disorder is very wet or extremely dry soils causing nutritional imbalances in the plant.

To control leaf necrosis, avoid very wet or dry soils and maintain optimum moisture content of the soil. It can also be controlled by providing moderate shading to small trees to reduce transpiration.

❑❑❑

Chapter-35

Persimmon

Scientific Name : *Diospyros kaki*

Family : Ebenaceae

Japanese persimmons were introduced into the United States from Japan by Admiral Perry who discovered the fruit growing on the coast of Southern Japan in 1851. It is regarded as the national fruit of Japan and several improved cultivars have been developed there. It is the one of the easiest tree fruits to grow in the home landscape and also the most beautiful especially in fall when the foliage turns to brilliant colours and huge orange fruits hang from the tree like ornaments.

Locally called "Japani Phal", or the Japanese fruit, persimmon looks like a bright red-orange tomato, and is full of subtle fragrances and rich in sugar, as well as vitamins (A, B and C). Persimmons are a good source of energy and dietary fiber. The wood is among the hardest known to man, being highly prized and desirable for wood carving by Japanese artists. The heartwood, the wood from the center of the tree makes decorative trims for furniture as well as other specialty items. Textile shuttles and driver golf club heads are made from the sapwood, the wood just under the bark of the persimmon tree.

In India, persimmon was introduced by European settlers in the early part of last century. Now, it is confined to small pockets as home trees or in orchards in Jammu and Kashmir, Coonoor in Tamil Nadu, Himachal Pradesh and hilly tracts of Uttar Pradesh and in northeastern India. The trees are deciduous and enter a rest period and complete their dormancy in the middle of February in India. No organized cultivation of this fruit has yet been undertaken in India but with the diversification in fruit culture, its cultivation is gaining importance and more areas are being opened for its cultivation. There are several problems which have limited the commercial cultivation of persimmon. The growers also face the problem of poor setting or heavy dropping of young fruits due to inadequate cultural knowledge resulting in poor returns. Other problems like calyx cavity and chilling injury have also to be checked. These problems have to be tackled at initial stages to promote its cultivation on large scale so that growing demand for this fruit undoubtedly could be fulfilled.

1. Fruit drop

Fruit drop is a common physiological problem for persimmons related to a number of factors such as lack of pollination, excessive fruit rot, water stress, uneven watering patterns, overgrown shoots, inadequate sunlight, varieties, age of tree, excessive nitrogen application and insect damage. Fruit drop is also caused by excessive vegetative growth. Persimmons do not need large amounts of fertilizers. Too much fertilization coupled with optimum soil moisture can produce excessive growth, thus causing more fruit drop than in a slower growing tree. Young trees drop more fruit, especially under stress than older trees.

Persimmons produce many blossoms and set up much more fruit than the tree could ever support. Due to excess burden on the tree its branches bend under the weight of the developing fruit taking shape of weeping willow. This excess fruit starts dropping to the ground sometimes leaving nothing on the tree. The fruit drop in persimmon occurs at various stages. The first wave of drop occurs in early June, initiating after petal fall and continues upto late July.

Thereafter, no fruit drop occurs in most of the varieties but in some varieties, late drop is also noted which is not equivalent to pre-harvest drop of fruits like mango, citrus and apples. It seems to be a unique feature of persimmon. The nutritional condition of trees affects the late drop. Some varieties are more prone to fruit drop when young but grow out of it with age. Various cultural practices such as ringing, blossom thinning and nitrogenous fertilizer applications reduce fruit drop.

2. Calyx cavity

Calyx cavity is a serious malady in persimmon characterized by a sparse space or cavity that occurs directly beneath the calyx of the fruit which becomes habitat for mealy bugs and fungal growth. Fruit with calyx-separation usually colours early with an uneven finish, leading to downgrading. The incidence of calyx cavity appears to be less on trees which have heavier crop loads and where fruits have been pollinated. Some varieties are more susceptible than others. Excessive fertilization causes calyx cavity in persimmons. Also deep, fertile and poor draining soils encourage this disorder. Similarly, the incidence is higher in areas with high autumn rainfall.

Therefore, to control this problem avoid excessive N and K fertilizers, especially close to harvest. Practice fruit thinning to enhance calyx growth and optimize pollination to produce more than 3 seeds per fruit. Choose sites with well drained soils for persimmon plantation.

3. Chilling injury

Chilling injury is a storage disorder of persimmon, the incidence and severity of which depend upon the cultivar, storage temperature and duration. The symptom development is fastest at 5-7°C. Upon transfer to higher temperature the severity of the symptoms such as flesh softening, browning and water soaked appearance, increases and renders the fruits unmarketable. It can be a major cause of deterioration during marketing after exposure to temperatures below 15°C. Exposure to ethylene at 1ppm or higher aggravates chilling symptoms.

It can be controlled by avoiding exposure of persimmons to temperature between 2°C and 15°C. The recommended storage and transport temperature for persimmons is 0°C so persimmons should be stored at this optimum temperature only. Also avoid exposure to ethylene above 1ppm throughout post-harvest handling of persimmons.

4. Skin russeting

Skin russeting is more prevalent in astringent cultivars such as Hachiya. It is characterised by concentric shallow rings around the fruit. The causes are not known, but it can be due to excessive nitrogen, irregular irrigation or high relative humidity during fruit ripening. Other possible causes may be damage due to birds or insects. To avoid skin russeting apply optimum nitrogen dose and give regular irrigation.

❑❑❑

Chapter-36

Pineapple

Scientific Name : *Ananas comosus*

Family : Bromeliaceae

Pineapple is the second most popular tropical fruit next to bananas. It is also known as Piña, Nanas and Ananas. Native to South America, it was named for its resemblance to a pine cone but it is neither a pine nor an apple. When Christopher Columbus made his second voyage to the Caribbean's in 1493, he and his crew ate the unusual fruit they found there. They thought the fruit looked like a pine cone, so they dubbed it the "Pine of the Indies." When they introduced it to the English later on, they added the word "apple" because they thought it should be associated with another delicious fruit that people enjoyed. Thus, the name "pineapple" came into existence. In their natural form, every variety of pineapple has a rough, diamond-pattern skin. Their tastes vary slightly, though they all basically have the same juicy, tart taste.

Pineapple usually contains vitamin C, B_1 and smaller amounts of B_2, B_3, B_5 and B_6. It is also an excellent source of manganese, copper, magnesium, potassium, sodium, beta-carotene, folic acid and dietary fiber but most importantly, they are a great source of an enzyme called Bromelin. It is a proteolytic enzyme which breaks down protein.

Pineapple juice can thus be used as a marinade and tenderizer for meat. Pineapple has minimal fat with no-cholesterol. In the Philippines, pineapple leaves are used as the source of a textile fiber called Pina.

The pineapple cultivation is becoming lucrative as the produce is very much in demand as a fresh fruit throughout India. Kerala's pineapple is considered the best in quality, sweetness and has good flavour. The pineapple cultivators are facing several problems related to quality such as sunscald, multiple crown or fasciation, black heart, flesh translucency, chilling injury, etc. Preventive measures taken for these disorders of pineapple will no doubt add revenue to the nation and growers.

1. Sunscald

Growers incur heavy losses in pineapple due to this malady. Sunscald is caused due to direct fall of sunrays on exposed area of the fruits making them unfit for consumption. The cells under the skin of exposed surface get damaged, due to which the peduncle bearing the fruit falls on the side exposed to the sun. The widely spaced plants are more prone to this disorder as compared to closely spaced plants. Thus, to control lodging or leaning, special care should be given. The fruits should be covered with dry straw or banana leaves or with its own leaves during the hot desiccating months of April-May.

This malady can be reduced considerably by adopting high density planting in pineapple.

2. Multiple crown or fasciation

Multiple crown or fasciation, a physiological disorder of pineapple occurs due to genetical as well as soil and environment factors. In normal cases, pineapple fruit bears a single crown however, in some cases it bears more than one crown resulting in broad top of the fruits. Sometimes the fruit get flattened and fasciated. Such fruits are corky and insipid and become unfit for canning. The incidence is more in Cayene group grown in fertile virgin soil of warm areas where more abnormal fruits occur as compared to crop grown in less fertile soil.

Causes

The factors associated with the fascination are:

i. High vigour of plants with more vegetative growth.

ii. High fertility of the soil.
iii. Warm weather.
iv. Advanced ratooning.

Control

The fasciation disorder can be avoided by following practices

a) Restricting the growth of the plant.
b) Avoiding advance ratooning.

3. Black heart or endogenous brown spot (EBS)

Black heart or endogenous brown spot, also known as internal browning is a serious disorder of pineapple. Symptoms are water-soaked, brown areas that begin as spots in the core area and enlarge to make the entire center brown in severe cases. Later these brown discoloured spots turns black and cover the entire centre of the fruit. Other symptoms are increased susceptibility to decay, wilting and discolouration of crown leaves.

Causes

EBS is usually associated with exposure of pineapples before or after harvest to chilling temperatures generally below 7°C (45°F) for one week or longer.

Control

Waxing is effective in reducing chilling injury symptoms. A heat treatment at 35°C (95°F) for one day is helpful to ameliorate EBS symptoms in pineapples transported at 7°C (45°F) by inhibiting activity of enzyme polyphenol oxidase and consequently tissue browning.

4. Flesh translucency or porosity

In this disorder fruit becomes more sensitive to the mechanical injury and the fruit pulp becomes translucent resulting in senescence. This condition begins before harvest and continues after it. A period of 2-3 months before harvest is critical for the development of such fruits. Translucency in fruits increases the fruit pH, TSS: acid ratio, weight and decreases the acidity. The maximum and minimum temperature preceding three months before harvest contributes to this malady. The incidence is high when these temperatures are lower i.e. 23°C and 15°C, respectively.

The severity of translucency decreases in fruits with large crown. The incidence can be lowered by waxing the fruits after harvest.

❑❑❑

Plum

Scientific Name : *Prunus salicina/domestica/americana et.*

Family : Rosaceae

A plum or gage is a stone fruit tree in the genus *Prunus*. The European plum is thought to have been discovered around two thousand years ago, originating in the area near the Caspian Sea, while Japanese plums actually originated in China. However, they derived their name from the country where much of their cultivation and development have occurred. Mature plum fruit may have a dusty-white coating that gives them a glaucous appearance and is easily rubbed off. This is an epicuticular wax coating and is known as "wax bloom". Plum fruit tastes sweet and/or tart; the skin may be particularly tart. It is juicy and can be eaten fresh or used in jam-making or other recipes. Plum juice can be fermented into plum wine; when distilled, this produces a brandy known as Slivovitz or Rakia in Eastern Europe. Dried plums (or prunes) are also sweet and juicy and contain several antioxidants. Large amounts of Vitamin C, K and A are present in plum besides these it also contains considerable amounts of thiamine, riboflavin and niacin whereas, vitamin E, vitamin B_6, folate and pantothenic acid are present in traces. It is enriched with minerals like potassium, magnesium, manganese, copper and phosphorus. While calcium, iron and zinc are present in minute quantities.

Generally plum is a temperate fruit but it is also cultivated in subtropics. Subtropical plums are cultivated on limited scale in the plains of north-western India. Although the fruit quality of plums produced in the subtropical climate is poor, still these fetch a high price in the market because of its early availability. It is available in the market at a time when the competition from other fruits is the least. Punjab is a leading state for cultivation of sub-tropical plum in India. Since plums are very delicate and perishable, they should be packed with care. Plums face some physiological problems such as heat spot due to high temperature and internal browning during cold storage.

1. Heat spot or Kelsey spot

This is a physiological problem in European plum which occurs due to high temperature and boron deficiency. Such spots deteriorate the quality of plum fruits. The disorder can be controlled by soil or foliar application of Borax @ 500 ppm in September to avoid the deficiency of boron. The high temperature can be controlled by white washing of stem to avoid heat spot in plum.

2. Internal breakdown or chilling injury

Internal browning is a limiting factor in the shipping of plum. It is characterized by flesh browning, flesh translucency, lack of juiciness in flesh which is due to leatheriness or mealiness. The other symptoms are formation of black pit cavity, flesh bleeding i.e. red pigment accumulation, failure to ripen and loss of flavour. These symptoms develop in plum and fresh prunes during ripening after a cold storage period. Thus, these symptoms are usually detected by consumers. Fruit stored within the killing temperature range of 2-8°C (36-46°F) are more susceptible to this problem. Internal browning can be controlled by temperature management. Storage below 0°C (32°F) but above the freezing point is beneficial to delay the symptoms and extend the market life.

❑❑❑

Chapter-38

Pomegranate

Scientific Name : *Punica granatum*

Family : Punicaceae

Fruit cracking

Pomegranate commonly known as 'anar' is an ancient favourite table fruit of tropical and sub-tropical regions of the world. The pomegranate tree is native from Iran to the Himalayas in northern India and has been cultivated since ancient times throughout the Mediterranean region of Asia, Africa and Europe. The word pomegranate comes from the Latin "*Pomum granatum*" which means "apple of many seeds." The Romans called it the *Punic apple*. Ancient Romans not only enjoyed the succulent flesh of this fruit but due to the high amount of tannic acid in the skins, they also used the skins in the process of tanning leather. Perhaps due to the fruit's princely blossom crown, it has gained distinction as a royal fruit. "*Pomum granatum*" or seeded apple was the name which was known in the middle ages. The French named their hand-tossed explosive a *grenade* after the seed-scattering properties of the pomegranate fruit.

The fruit is symbolic of plenty and very much liked for its cool, refreshing juice and valued for its medicinal properties. The juice of pomegranate is believed to be good for leprosy patients. The grains of the fruit are also eaten fresh in most of the hot countries and are used as condiment. The bark and rind of the fruits are commonly used in dysentery and diarrhoea. Dried seeds of pomegranate seeds with pulp are available as 'Anardana'. The trunk bark contains 10 to 25 per cent tannin and was formerly important in the production of Morocco leather. Both the rind and the flowers yield dyes for textiles. The pale-yellow wood is very hard and available only in small dimensions is used for walking-sticks and in woodcrafts. In some countries such as Iran the juice is a very popular beverage.

Pomegranate is not a common as it cost slightly more than the common fruits. But, the health benefits of its red juice are tremendous. Inside a pomegranate is about 700-800 tightly packed seed casings called arils that are deep red in colour when nicely ripe. The taste of the juice basically is sweet, sour or tangy. Pomegranates have very high content of punicalagins, a potent anti-oxidant component found to be responsible for its superior health benefits. Amazingly, researchers indicate that the capacity of anti-oxidant in this fruit is two or three times higher than that of red wine and green tea. They are also a good source of vitamin B complex (riboflavin, thiamin and niacin), vitamin C and minerals like calcium and phosphorus. These characters in pomegranate cause a powerful synergy that prevents and reverses many diseases.

The versatile adaptability, hardy nature, steady but high yields, better keeping quality, fine table and therapeutic values indicate the avenues for increasing the area under pomegranate in India. However, it is confronted with some physiological disorders which deteriorate its quality hindering the production of superior end product.

1. Internal breakdown

Mature pomegranate fruits are susceptible to internal breakdown i.e. disintegration of arils or blackening of arils. This serious malady may cause 50-60 per cent loss of produce. The disorder cannot be identified externally whereas, the arils become soft, light creamy-brown to dark blackish brown and unfit for consumption. The incidence of internal breakdown occurs 90 days after anthesis. Its intensity increases if the fruits are left on the tree for 140 days onwards. Both evergreen and deciduous cultivars are prone to internal breakdown but the incidence is more in Ambe-bahar crop. It increases with increase in weight. The TSS, acidity, ascorbic acid, reducing sugars, calcium, phosphorus and enzyme catalase are reduced whereas non reducing sugars, starch, tannis, nitrogen, potassium, magnesium, boron and enzyme polyphenol oxidase and peroxidase increases in the affected arils compared with the healthy ones. No insect or organism is associated with this malady.

To avoid the incidence to some extent, the fruit should be harvested at 120-135 days after fruit set.

2. Fruit cracking

Fruit cracking or splitting is a major physiological disorder of pomegranate in India. Splitting is of different types, however, longitudinal and radial are common types. The longitudinal cracking is more prevalent where the crack originates at the naval or stylar end of the fruits. However, in some fruits, radial cracking is also seen where split originates between the ends. Fruit cracking, a serious problem, is more intense under dry conditions of the arid zone. The cracked fruits are breeding grounds for insects, harmful bacteria and fungi making fruits unfit for marketing and human consumption. These fruits are also liable to rot qualitatively. The Mrig bahar crop is more susceptible to cracking than the crop of other bahars.

Causes

Large number of factors are found associated with fruit cracking. These are :

i. Sudden rains or irrigation followed by prolonged drought.

ii. Fluctuating soil moisture level.

iii. Variation in temperature and relative humidity.

iv. High nitrogen.

v. Deficiency of calcium and boron.

vi. Hormonal imbalances

Main cause of this malady is the wide variation in moisture content of the soil as well as in the humidity of area due to monsoon. Fully developed pomegranate fruits crack due to moisture imbalance as they are very sensitive to variation in soil moisture and also to day and night atmospheric moisture deficit. Prolonged drought causes hardening of peel. If this is followed by heavy irrigations or rains, the pulp grows and the peel cracks as it looses its elasticity.

The cracking of a fruit is mostly due to irregular water supply to the trees. As Ambe- bahar crop is regularly irrigated so it does not crack badly. The best treatment is to give regular irrigations to the crop taking care that at no stage there is a scarcity of water. In case of Mrig-bahar crop, the splitting of fruits cannot be controlled altogether as the variation in humidity cannot control cracking. However, it can be minimized if the plants are regularly irrigated wherever there is break in rain.

Control

a. Cracking is correlated with rind thickness. Cultivars like Karkai, Bedana, Khog, Guleshah and Jalore Seedless are comparatively tolerant to fruit cracking. So plant only those varieties which are less prone to fruit cracking.

b. Irrigate field regularly during summer months so as to maintain soil moisture.

c. Plant wind breaks around the pomegranate plantation.

d. Apply some mulch during summer months to conserve soil moisture.

e. Spray 100 ppm NAA and 100-200 ppm GA_3 during the fruit growth period.

f. Spray 0.8 per cent Borax to check fruit cracking.

3. Husk scald

Scald is a brown superficial discolouration restricted to the husk. At advanced stages, the scalded areas became mouldy. The scald symptoms become evident after 8 weeks of storage at 2^0 C. Symptoms appear earlier at higher storage temperatures. Husk scald development of pomegranate may be due to phenolic oxidation. As late harvested fruit is less susceptible than earlier harvested fruit, so harvest the fruit few days late.

4. Chilling injury

Chilling injury can be a major cause of deterioration of pomegranates during marketing following exposure to temperatures below 5°C (41°F) during storage and transport for longer than 4 weeks. External symptoms include brown discolouration of the skin, pitting and increased susceptibility to decay. Internal symptoms include brown discolouration of the white segments separating the arils and pale colour i.e. loss of red colour of the arils. External and internal browning is related to oxidation of phenolics by polyphenol oxidase.

The incidence and severity of chilling injury depend upon storage temperature and duration. Their severity increases with lower temperatures and longer durations of chilling exposure. Pomegranates are susceptible to chilling injury if stored longer than one month at temperatures between -3°C and 5°C. Upon transfer to higher temperatures, respiration and ethylene production rates increase and chilling injury symptoms appear. Avoiding exposure of pomegranates to temperatures below 5°C can control this disorder.

❑❑❑

Chapter-39

Raspberry

Scientific Name : *Rubus* spp.

Family : Rosaceae

Raspberries are grown for the fresh fruit market and for commercial processing. There are many species of raspberry but the name originally referred with the red fruit to the European species *Rubus idaeus*. The English name comes from an old English term "raspis", which refers to the slightly hairy or "rasping" surface of the fruit. The raspberry can be of any colour from orange, pink, red, purple to black. *Rubus albescens* also known as Mysore raspberry is native to India. It is also called Snowpeaks, Ceylon or Hill raspberry. The fruit turns from red to purple and black when ripe. The Mysore berry has been in great demand by hotel chefs. The fruits are enjoyed fresh, alone or served with sugar and cream or ice cream. They are excellent for making pie, tarts, jam and jelly. The fresh fruits can be quick-frozen for future use. Raspberries are a valuable source of Vitamin C, niacin, riboflavin, manganese, potassium and dietary fiber. Their juice is said to be good for the heart, and the leaves have long been used for their beneficial effects during childbirth.

Mysore raspberries should be picked for market when they are 3/4 purple-black. These should be harvested daily during peak periods. Fully ripe purple berries have a shorter shelf life and harvested for culinary use or frozen for future use. Loss of marketable fruit can be as high as 30% in some raspberry growing areas due to poor handling during the harvest and improper post-harvest care. High temperature is a limiting factor in raspberry cultivation as it results in disorder such as white drupelet due to sun exposure resulting in poor quality fruit.

1. White drupelets

White drupelet disorder is a tan to white discolouration of individual drupelets on raspberry fruits. About 75 to 85 drupelets mature into the 'aggregate' fruit of raspberry. Sometime these drupelets are white in colour i.e. drupelets develop normally but are lacking in colour, usually making the berries unacceptable for fresh market but still usable for processing. This can be caused due to sun exposure by high temperatures (above 40°C). Berries with full exposure to direct afternoon sun appear most susceptible to white drupelet disorder. Injury has been observed to rapidly increase as berries move from green to white to pink stages during ripening. However, high temperatures also appear to be involved as berries shaded by canopy in May also develop white drupelet disorder, which may become quite prevalent during hot growing seasons.

Control

To prevent white drupelets following cultural practices should be followed:

a. Avoid planting on sites with strong hot summer winds.

b. Use overhead system for evaporative cooling twice a day for about 15 minutes between 10 am to 3 pm to prevent sunscald.

c. Shading may help minimize the occurrence of white drupelet disorder by reducing associated UV radiation.

d. Covering the fruits with suitable covers such as black fabric or white polyester at least two weeks before harvest also miminize the tanning or discolouration in raspberries.

❑❑❑

Sapota

Scientific Name : *Manilkara achras*

Family : Sapotaceae

Sapota or sapodilla popularly known as *Chiku* is native to tropical America where it is mainly grown for its chickle, "the gutta parcha" extracted from its latex from stems. It is used as a base material in chewing gum and in some other industrial uses. However, in India it is cultivated for its delicious sweet fruits. The fruit is fleshy berry, variable in shape, size and weight. The skin is thin, rusty brown somewhat scurfy looking like Irish potato, and the pulp soft, melting, crumbling with a sandy or granular texture with 1-5 hard black seeds. The fruit has an exceptionally sweet malty flavour. Many believe the flavour bears a striking resemblance to caramel. The unripe fruit is hard to touch and contains high amounts of saponin, which has astringent properties similar to tannin, drying out the mouth. The fruit is a good source of digestible sugar (15-20%) and an appreciable source of protein, fat, fibre and minerals like calcium, phosphorus and iron. Sapota pulp is used for making sweets and halwas. It is also an ingredient of fruit salads and milk shakes.

Sapota is a popular fruit crop in Gujarat, Maharashtra, Karnataka, Tamil Nadu, Andhra Pradesh and Kerala. The fruit is mostly consumed indigenously and export constitutes only a very minor fraction of production. Low volume export of sapota is due to non-ideal post harvest practices, transport procedures, lack of proper storage facilities and outdated handling practices. Sapota fruit is highly perishable and is also sensitive to cold storage. Therefore, bulk of the produce is used for table purpose and is handled at ambient climatic conditions causing considerable post harvest losses. Due to mishandling of produce about 25-40 per cent is being wasted. Commercial processing is negligible due to the sensitivity of the fruit to heat, changing the flavour and colour of the pulp. The major physiological disorders of sapota are wilt, heat and chilling injury.

1. Wilt

Wilt or die back is common where sapota cultivation is being extended to traditionally rice growing regions. Due to anaerobic conditions in monsoon and post monsoon season in such areas, wilt is of common appearance. This can be controlled by effective drainage facility before planting.

2. Oblong fruits

High temperature and rainfall during flowering cause oblongation of fruits. The shape of fruit is related with number of seeds in it which depend on conditions for pollination at anthesis. Therefore, cultivation of sapota in areas with extreme summer temperature should be avoided.

3. Fruit depression

Development of a depression or furrow towards the calyx-end usually appears immediately after heavy rainfall and is aggravated by high intensity of irrigation. The fruits do not develop into their normal shape. Avoiding over-irrigation can check this malady.

4. Corkiness

Corkiness develops during winter when the fruits are exposed to intense sunlight. This is probably due to killing of hydrolyzing enzymes by alternating moisture accumulation and heating of fruit surface in winter. The fruits do not ripen uniformly due to corkiness. Corkiness can be avoided if the trees are made to grow more vigorous.

5. Chilling injury

Sapodilla fruits are highly susceptible to chilling injury. Storage of fruit at 6 to 10 °C causes irreversible damage and results in poor flavour. It also occurs when fruits are stored for 21 days at 10°C. However, waxing of fruits checks this problem.

❑❑❑

Chapter-41

Strawberry

Scientific Name : *Fragaria x ananassa*

Family : Rosaceae

Strawberries one of the best fruits of the world are native to North America. Strawberry fruits are aggregates made up of several small fruits, each with one seed called an achene. The flesh of the strawberry is actually an enlarged receptacle, non-reproductive material. The word *strawberry* comes from the old English *streawberige*, most likely because the plant sends out runners which could be linked to pieces of straw. The name Strawberry was also derived from the berries that are "strewn" about on the plants, and "strewn berry" eventually became "Strawberry". Strawberry is a delicious fruit taken fresh in several ways. It also makes excellent ice cream and jam on account of its rich aroma.

The berries are non-fat and low in calories. The strawberries provide an excellent source of vitamins (B, C and K), minerals (copper, magnesium, potassium and manganese), folic acid and omega-3 fatty acids. The strawberries have been used in medicines since time immemorial. They have been used for sunburn and discolored teeth. Strawberries are great in pie, jam, ice cream, cake and as fruit leather. They are also fabulous just with a sprinkling of sugar or a splash of cream.

In India it is generally cultivated in the hills. Its main center of cultivation is Nainital and Dehradun in Uttar Pradesh, Mahabaleshwar (Maharashtra), Kashmir Valley, Bangalore and Kalimpong (West Bengal). In recent years, strawberry is being cultivated successfully in plains of Maharashtra around Pune, Nashik and Sangali towns. It is also grown in Punjab with certain environmental modifications. Strawberries are highly perishable and hence a great deal of care is required in harvesting and handling as well as its marketing also requires to be organized carefully. Usually the fruit is picked in the early morning and sent to the market in the afternoon of the same day or is picked in the late afternoon, stored overnight in a cool place, and sent to market the following morning. The disorders discussed below are bottlenecks in strawberry cultivation.

1. Albinism

It is most serious disorder of strawberry in which fruits get blotted developing white or pink discolouration on the surface while the pulp gives pale colour. The affected fruits having poor flavour, more acidity and less firmness are not acceptable by the consumer. Albino fruits are often damaged during harvesting and are susceptible to *Botrytis* infection and decay during storage. Though the fruits develop normal but they do not ripe uniformly. The fruits are little bit waxy in appearance and fetch poor price in the market. Such fruits are liable to severe damage during harvesting. In storage, fruits become susceptible to fruit rot.

Causes

The main causes of albinism are discussed below.

i. High density planting.

ii. Excessive fertilization.

iii. Warm weather and overcast sky during peak fruit production in the preceding season.

iv. It develops more in sandy and low pH soils.

v. High level of N, P and K content in the soil enhances this disorder.

vi. Severity increase in closed tunnels with less ventilation.

vii. Albinism incidence is more with black film mulch than white film.

viii. It may be due to varietal feature as cultivar like Elasanta is worst affected.

Control

Following control measures help to control albinism.

a) Select suitable variety.

b) Avoid dense planting.

c) Avoid excessive fertilization with N and K.

d) Use hay or paddy straw as mulch material instead of polythene mulch.

2. Fasciation

Fasciation is a major problem in strawberry cultivation which is characterized by abnormal flattening of fruits and their stems. The fruit gets enlarged and plant gives witch's broom appearance. There is little or no runner production. Even the yield is greatly affected as such plants produce little marketable fruits. In severe cases the flower buds broadens and no fleshy fruit develops in the spring rendering the plants completely barren. If berries develop, they develop into a typical shape of cock's comb.

Causes

Fasciation is caused due to following factors.

i. Unfavourable growth conditions when days become too short for normal development of fruit.

ii. Varietal character as some varieties such as Missionary fasciates more and some like Blackmore never fasciate.

iii. Infestation due to insect and pests.

iv. Infestation due to diseases.

v. Aggravation of humidity and frost.

vi. Lack of pollination.

Control

To avoid fasciation below mentioned practices should be followed.

a) There should be provision of suitable pollinator.

b) Proper management of strawberry plantation should be done.

3. Fruit malformation

It is commonly observed in strawberry cultivation that some plants produce malformed fruits. The shape of the berries gets distorted. It is seen that the primary and secondary flowers produce more malformed fruits. This is attributed due to the undeveloped achenes at the distal end of the receptacle than the tertiary and quaternary receptacle.

Causes

Following factors are associated with fruit malformation in strawberry.

i. Excessive nitrogen fertilization
ii. Planting of more vigorous plants
iii. Insufficient pollination
iv. Lack of growth promoting substances

Control

a) Avoid excessive application of N fertilizers
b) Plant young and less vigorous plants
c) Provide adequate pollinizers.
d) Also provide adequate honey bee hives in the commercial planting

4. Phyllody

This disorder is characterized by flattening of flowers and fruits. The petals of affected flowers get thickened and develop poor colour. Firstly, the opening of flowers is affected and then the fruit set. As it is difficult to control phyllody so it is better to uproot and burn the infected strawberry plants.

❑❑❑

Chapter-42

Walnut

Scientific Name : *Juglans regia* L.

Family : Juglandaceae

The English walnut (*Juglans regia*) is a native of Western and Central Asia. The nut was the food of the royalty and was introduced by the Greeks in Rome, where it became *Juglans*, a name derived from *Jovis* and *glans*, meaning Jupiter's Acorn or the 'Nut of the Gods'. The English walnut tree, also known as the Common, Persian or Royal walnut tree, is an ancient tree with many uses, most notably for its use as timber and for the walnut fruit. The common walnut tree has been used as a food, beauty aid, medicine and natural health supplement. In China the tree is used horticulturally in classical garden design.

As a food the walnut fruit is used in cakes and salads as a flavouring agent. Walnuts are highly nutritious as 100g shelled walnuts provide 15.2 g protein, 65.2 g fat, 13.7 g carbohydrates, 6.7 g dietary fiber, 0.34 mg thiamine, 0.54 mg Vitamin B_6 and minerals like manganese, copper, magnesium, zinc and phosphorus. Walnut wood is of very high quality which is used for making furniture and gunstocks. The red coloured wood of the common walnut tree was, and still is today, considered to be luxurious. Walnut wood is often found in the fittings of luxury cars. Walnut ink, made by boiling the whole fruit or letting it oxidize is dark brown in colour and darkens as it oxidizes. It can also be used to stain wood. The walnut oil is used in culinary dishes, in beauty preparations for hair and skin and is also used as therapeutic oil in aromatherapy and massage.

The state of Jammu and Kashmir in the northern region of India is the country's major walnut producing zone. The English walnut tree is capable of producing more nuts than any other walnut tree. The tree is hardy and can tolerate cold. The problems like oil rancidity can adversely affect the quality produce.

1. Oil rancidity

The most serious postharvest physiological disorder that affects walnut quality is oil rancidity. The problem appears to be caused by poor seed storage conditions; elevated temperature and relative humidity, failure to use controlled atmosphere and imbalance in O_2 concentration. The seed kernels are enclosed in a brown seed coat which contains antioxidants. The antioxidants protect the oil-rich seed from atmospheric oxygen so preventing rancidity (oxidative rancidity). Walnuts have the best colour and flavour when their water content is 2 - 8 per cent. Higher water contents reduce storage life and increases the risk of rancidity. Rancidity caused by oxidative fat cleavage is particularly noticeable in the case of shelled walnuts, because of exposure to atmospheric oxygen. It is therefore absolutely essential to store walnuts in the dark and to protect them from oxygen and since otherwise they become brown-coloured and develops a rancid odour and taste.

2. Winter sunscald

This type of injury occurs when the sun warms tree bark during the day and then the bark rapidly cools after sunset. These abrupt fluctuations may kill the inner bark. Young trees are most susceptible to winter sunscald. Wrapping trunks of susceptible trees with protective "tree wrap" is the most effective way to minimize this type of winter injury.

❑❑❑

References

1. Bal, J.S. (1997). Fruit Growing. Kalyani Publishers, Ludhiana, India.
2. Bose, T.K., Mitra, S.K. and Sanyal, D. (2001). Fruits : Tropical and Sub-tropical. Vol. I. Naya Udyog, Kolkata, India.
3. Chadha, K.L. (2001). Handbook of Horticulture, ICAR, New Delhi, India.
4. Chattopadhyay, T.K. (1998). A Textbook on Pomology. Vol. I. Kalyani Publishers, Ludhiana, India.
5. Chattopadhyay, T.K. (1998). A Textbook on Pomology. Vol. II. Kalyani Publishers, Ludhiana, India.
6. Chattopadhyay, T.K. (1998). A Textbook on Pomology. Vol. III. Kalyani Publishers, Ludhiana, India.
7. Chattopadhyay, T.K. (1998). A Textbook on Pomology. Vol. IV. Kalyani Publishers, Ludhiana, India.
8. Gardner, V.R., Bradford, F.C. and Hooker, H.G. (1939). The Fundamentals of Fruit Production. McGraw Hill Book Company, Inc, New York, USA.

9. Nakasone, H.Y. and Paull, R.E. (1998). Tropical Fruits. CABI Publishing, New York, USA.

10. Salaria, A.S. (1999). Horticulture : at a glance. Jain Brothers, New Delhi, India.

11. Samson, J.A. (1980). Tropical Fruits. Longman, London, U.K.

12. Chadha, T.R. (2003). A Textbook on Temperate Fruits. Indian Council of Agricultural Research, New Delhi, India.

13. Childers, N.F., Morris, J.R. and Sibbett, G.S. (1995). Modern Fruit Science. Hort. Publication, Gainesville, Florida, USA.

14. Baritelle A.L. and Hyde, G.M. (2001) Commodity conditioning to reduce impact bruising. *Postharv. Biol. Technol.* 21(3): 331-339.

15. Ruiz Altisent, M. (1991) Damage mechanisms in the handling of fruits. In 'Progress in Agricultural Physics and Engineering' (Ed Matthews J.) CAB International, 231-257.

16. Jindal, K.K. and Gautam, D.R. (2003). Enhancement of temperate fruit production. Compendium of summer school on enhancement of temperate fruit production in changing climate held at UHF, Solan. w.e.f. September 19 to October 9, 2001.

17. Mitra, S.K., Rathore, D.S. and Bose, T.K. (1991). Temperate fruits. Allied Publishers, Kolkata, India.

18. Samaddar, H.N. (2001). Commercial production of Horticultural crops. Naya Udyog, Kolkata, India.

19. Singh, A. (1996). Physiology of fruit production. Kalyani Publishers, Ludhiana, India.

20. Bangerth, F. (1979). Calcium related physiological disorders of plants. *Annual Review of Phytopathology,* 17: 97-122.

21. Crisosto, C.H., Johnson, R.S., DeJong, T. and Day, K.R. (1977). Orchard factors affecting postharvest stone fruit quality. *HortScience,* 32: 820-823.

22. Sharma, R.R. (2006). Fruit production. Problems and Solutions. International Book Distributing Co., Lucknow, U.P., India. pp. 301-339.

23. Saltveit, M. E., Jr. and Morris, L. L. (1990). Overview on chilling injury of horticultural crops. *In:* Chilling injury of horticultural crops. (C. Y. Wang, ed.). pp. 313 CRC Press, Inc., Boca Raton, Florida, USA.

24. Battey, N.H. (1990). Calcium deficiency disorders of fruits and vegetables. *Postharvest News and Information,* 1(1): 23-27.

25. Bramlage, W.J. (1982). Chilling injury of crops of temperate origin. *HortScience,* 17: 165-168.

26. Couey, H.M. (1982). Chilling injury of crops of tropical and subtropical origin. *HortScience,* 17: 162-165.

27. Ferguson, I.B. (1980). Mineral composition of fruit and vegetables in relation to storage life. *CSIRO Food Res. Quart.,* 40(3-4): 94-100.

28. Graham, D. and Patterson, B.D. (1982). Responses of plants to low, nonfreezing temperatures: proteins, metabolism and acclimation. *Annu. Rev. Plant Physiol.,* 33: 347-372.

29. Grierson, W., Soule, J. and Kawada, K. (1982). Beneficial aspects of physiological stress. *Hort. Rev.,* 4: 247-271.

30. Jackman, R.L., Yada, R.Y., Marangoni, A., Parkin, K.L. and Stanley, D.W. (1989). Chilling injury: A review of quality aspects. *J. Food Qual.,* 11: 253-279.

31. Kader, A.A., J.M. Lyons, and L.L. Morris. (1974). Quality and postharvest responses of vegetables to pre-harvest field temperature. *HortScience,* 9: 523-527.

32. Klein, J.D. and S. Lurie. (1991). Postharvest heat treatment and fruit quality. *Postharvest News and Information,* 2: 15-19.

33. Klein, J.D. and S. Lurie. (1992). Heat treatments for improved postharvest quality of horticultural crops. *Hort. Technol.,* 2: 316-320.

34. Lurie, S. (1998a). Postharvest heat treatments of horticultural crops. *Hort. Rev.,* 22: 91-121.

35. Lurie, S. (1998b). Postharvest heat treatments. *Postharvest. Biol. Technol.,* 14: 257-269.

36. Lyons, J.M. (1973). Chilling injury in plants. *Annu. Rev. Plant Physiol.,* 24: 445-466.

37. Morris, L.L. (1982). Chilling injury of horticultural crops: An overview. *HortScience,* 17: 161-162.

38. Opara, L.U., Studman, C.J. and Banks, N.H. (1997). Fruit skin splitting and cracking. *Hort. Rev.,* 19: 217-262.

39. Parkin, K.L., Marangoni, A., Jackman, R.L., Yada, R.Y. and Stanley, D.W. (1989). Chilling injury: A review of possible mechanisms. *J. Food Biochem.*, 13: 127-153.

40. Shear, C.B. (1975). Calcium-related disorders of fruits and vegetables. *HortScience,* 10: 361-365.

41. Smagula, J.M. and Bramlage, W.J. (1977). Acetaldehyde accumulation. Is it a cause of physiological deterioration of fruits. *HortScience,* 12: 200-203.

42. Wang, C.Y. (1982). Physiological and biochemical responses of plants to chilling stress. *HortScience,* 17:173-186.

43. Wang, C.Y. (1993). Approaches to reduce chilling injury of fruits and vegetables. *Hort. Rev.,* 15: 63-95.

44. Wang, C.Y. (1994). Chilling injury of tropical horticultural commodities. *HortScience,* 29: 986-988.

45. Wolk, W.D. and Herner, R.C. (1982). Chilling injury of germinating seeds and seedlings. *HortScience,* 17: 169-173.

46. Whiteman, T.M. (1957). Freezing points of fruits, vegetables and florist stocks. *USDA Mktg. Res. Rept.,* 196, pp. 32.

47. Wismer, W.V., Marangoni, A.G. and Yada, R.Y. (1995). Low-temperature sweetening in roots and tubers. *Hort. Rev.,* 17: 203-231.

48. Woolf, A.B. and Ferguson, I.B. (2000). Postharvest responses to high fruit temperatures in the field. *Postharv. Biol. Technol.,* 21: 7-20.

49. Yang, C.Y. (1991). Reduction of chilling injury in fruits and vegetables. *Postharvest News and Information,* 2: 165-168.

50. Hardenburg, R.E., Watada, A.E. and Wang, C.Y. (1986). The commercial storage of fruits, vegetables, and florist and nursery stocks. *USDA Agric.* Hb. No. 66.

51. Kays, S.J. (1991). Postharvest physiology and handling of perishable plant products. Van Nostrand Reinhold, New York, 532 p.

52. Ryall, A.L. and Lipton, W.J. (1979). Handling, transportation and storage of fruits and vegetables. In: *Vegetables and melon, Vol. 1,* (2nd ed). Avi Publ. Co., Westport, CT, 588 p.

53. Ryall, A.L. and Pentzer, W.T. (1982). Handling, transportation and storage of fruits and vegetables. In: *Fruits & tree nuts, Vol. 2,* (2nd ed). Avi Publ. Co., Westport, CT, 610 p.

54. Kader, A.A. (1985). Ethylene-induced senescence and physiological disorders in harvested horticultural crops. *HortScience,* 20: 54-57.

55. Tindall, H.D. and Proctor, F.J. (1980). Loss prevention of horticultural crops in the tropics. *Prog. Food Nutr. Sci.,* 4(3-4): 25-40.

56. Anonymous. (2003). Cultivating subtropical crops. ARC-Institute for Tropical and Subtropical Crops, Department of Agriculture, Directorate Agricultural Information Services. pp. 153.

57. De Villiers, E.A. (1998). The cultivation of Mangoes. ARC-Institute for Tropical and Subtropical Crops. Nelspruit, pp. 216.

58. De Villiers, E.A. (1999). The cultivation of Papaya. ARC-Institute for Tropical and Subtropical Crops. Nelspruit, pp. 98.

59. De Villiers, E.A. (2001). The cultivation of Avocado. ARC-Institute for Tropical and Subtropical Crops. Nelspruit, pp. 368.

60. De Villiers, E.A. (2002). The cultivation of Litchi. ARC-Institute for Tropical and Subtropical Crops. Nelspruit, pp. 213.

61. Du Plessis, S.F. (1993). Irrigation. *In: The Cultivation of Macadamias.* Institute for Tropical and Subtropical Crops, Bulletin 426. Aurora Printers, Pretoria, pp. 66.

62. De Villiers, E.A. and Joubert, P.H. (2003). The cultivation of Macadamia. ARC- Institute for Tropical and Subtropical Crops. Nelspruit, pp. 198.

63. Jackson, D.I. and Looney, N.E. (1999). Temperate and Subtropical Fruit Production. 2nd Eds. CABI Publishing, Wallingford, UK.

64. Linsley-Noakes, G.C. (1989a). Blueberry production in South Africa. *FFTRI Bulletin,* No.583.

65. Linsley-Noakes, G.C. (1989b). Raspberry production in South Africa. *FFTRI Bulletin,* No.586.

66. O'Hare, P. (1994). Growing papaws in South Queensland. The State of Queensland, Department of Primary Industries, Brisbane, pp. 41.

67. Smith, J.M.B. (1998). Handbook for Agricultural Advisors in KwaZulu-Natal. KZN Department of Agriculture & Environmental Affairs.

68. Vock, N. (1991). Growing strawberries in Queensland. The State of Queensland, Department of Primary Industries, Brisbane, pp. 43.

69. Zwahlen, K.G., Jackson, D.C., Erasmus, G.M.M and Myburgh, J. (1989). The cultivation of Cherries in the Eastern Free State. *Farming in South Africa, Cherries* C1/1989.

70. Coates, L., Cooke, A., Persley, D., Beattie, B., Wade, N. and Ridgeway, R. (1995). Postharvest diseases of horticultural produce, Volume 2: Tropical Fruit. DPI, Queensland.

71. Bohlmann, T.E. (1962). Why does fruits crack? *Fmg. S. Afr.*, 38 (7): 12-13.

AONLA

72. Allemullah, M. and Ram, S. (1990). Causes of low fruit set and heavy fruit drop in Indian gooseberry (*Emblica officinalis* Gaertn.). *Indian J. Hort.*, 47: 270-277.

73. Bajpai, P.N. and Shukla, H.S. (1990). Aonla. *In*: Bose, T.K. and Mitra, S.K. (eds). *Fruits: Tropical and Sub-tropical.* Naya Prakash, Kolkata, India. pp. 757-767.

74. Morton, J.F. (1987). Emblic. *In:* Morton, J.F. (eds.). *Fruits of warm climates.* Creative Resource System Incorporated, Winterville, U.S.A. pp. 213-217.

75. Pathak, R.K., Pandey, D., Haseeb, M. and Tandon, D.K. (2003). The aonla. *Bulletin CISH,* Lucknow, India.

76. Pathak, R.K. and Srivastav, A.K. (1994). Flowering and fruit set studies in Indian gooseberry. *Proc. Indian Hort. Congress*, Kyoto, Japan.

77. Shukla, A.K., Pathak, R.K, Tewari, R.P. and Nath, V. (2000). Influence of irrigation and mulching on plant growth and leaf nutrient status of aonla under sodic soil. *J. Applied Hort.*, 2(1): 37-38.

78. Tewari, D.N., Kumar, K. and Tripathi, A. (2001). Amla. Ocean Books Pvt. Ltd., New Delhi, India.

79. Dashora, L.K., Lakhawat, S.S. and Arvindakshan, K. (2005). Effect of foliar spray of zinc and boron on yield and quality of aonla. *Nat. Sem. Commercialization of Hort. in Non-traditional areas.* pp 85. C.I.A.H. Bikaner, India.

80. Singh, J.K., Parsad, J. and Singh, H.K. (2007). Effect of micro-nutrients and plant growth regulators on yield and physico-

chemical characteristics of aonla fruits in cv. Narendra Aonla-10. *Indian J. Hort.*, 64(2): 216-218.

APPLE

81. Taylor, B.K. and Knight, J.N. (1986). Russetting and cracking of apple fruit and their control with plant growth regulators. *Acta Hort.*, 179: 319-320.

82. Eccher, T. (1983). Control of russetting of 'Golden Delicious' apple by growth regulator treatments. *Acta Hort.*, 137: 375-382.

83. Byers, R.E., Yoder, K.S. and Mattus, G.E. (1983). Reduction in russetting of 'Golden Delicious' apple with 2,4,5-T and other compounds. *HortScience*, 18: 63-65.

84. Bowen, J.H. and Watkins, C.B. (1997). Fruit maturity, carbohydrate and mineral content relationships with watercore in 'Fuji' apples. *Postharvest Biol. Technol.*, 11: 31–38.

85. Bramlage, W.J. and Weis, S.A. (1997). Effects of temperature, light and rainfall on superficial scald susceptibility in apples. *HortScience.*, 32: 808–811.

86. Faust, M., Shear, C.B. and Williams, M.W. (1969). Disorders of carbohydrate metabolism of apples. *Bot. Rev.*, 35: 169– 194.

87. Ferguson, I.B. and Watkins, C.B. (1989). Bitter pit in apple fruit. *Hortic. Rev.*, 11: 289–355.

88. Ferguson, I.B. and Watkins, C.B. (1992). Crop load affects mineral concentrations and incidence of bitter pit in 'Cox's Orange Pippin' apple fruit. *J. Amer. Soc. Hort. Sci.*, 117: 373–376.

89. Marlow, G.C. and Loescher, W.H. (1984). Watercore. *Hortic. Rev.*, 6: 189–251.

90. Yamada, H., Ohmura, H., Ara, C. and Terui, M., (1994). Effect of preharvest fruit temperature on ripening, sugars, and occurrence of watercore in apples. *J. Amer. Soc. Hortic. Sci.*, 119: 208–1214.

91. Hatch, A.H. (1975). The influence of mineral nutrition and fungicides on russetting of 'Goldspur' apple fruit. *J. Amer. Soc. Hort. Sci.*, 100: 52-55.

92. Taylor, B.K. (1986). Effect of gibberellin sprays on fruit russet and tree performance of Golden Delicious apple. *J. Hort. Sci.*, 53: 167-169.

93. Faust, M. and Shear, C.B. (1968). Corking disorders of apples. A physiological and botanical review. *Bot Rev.*, 34: 411-469.

94. Faust, M. and Shear, C.B. (1972). The effect of calcium on the respiration of apples. *J. Amer. Soc. Hort. Sci.*, 97: 437-439.

95. Byers, R.E. (1993). Controlling growth of bearing apple trees with ethephon. *HortScience*, 28(11):1103-1105.

96. Dennis, Jr., F.G. and J.C. Neilsen. (1999). Physiological factors affecting biennial bearing in tree fruit: the role of seeds in apple. *Hort. Tech.*, 9(3):317-322.

97. Meador, D.B. and B.H. Taylor. (1987). Effect of early season foliar sprays of GA_{4+7} on russeting and return bloom of 'Golden Delicious' apple. *HortScience*, 22(3):412-415.

98. Singh, L.B. (1948a). Studies in biennial bearing II: A review of the literature. *J. Hort. Sci.*, 24:45-65.

99. Singh, L.B. (1948b). Studies in biennial bearing III: Growth studies in "on" and "off" year trees. *J. Hort. Sci.*, 24:123-154.

100 Singh, L.B. (1948c). Studies in biennial bearing IV: Bud-rubbing, blossom-thinning, and defoliation as possible control measures. *J. Hort. Sci.*, 24:159-177.

101 Matthais, J.P. (1995). Factors contributing to internal breakdown of Fuji apples. *Tree Fruit Postharvest J.*, 6:3-4.

102 Ingle, M. and M. C. D'Souza. (1989). Physiology and control of superficial scald of apples: a review. *HortScience,* 24:28-31.

103 Hall, E.G. and Scott, K.J. (1989). Pome Fruit. *In:* Beattie, B.B., McGlasson, W.B., and Wade, N.L. (eds.) *Postharvest Diseases of Horticultural Produce.* Vol. 1. Temperate Fruit. CSIRO, Australia.

104 Jones, A.L. and Sutton, T.B. (1984). Diseases of Tree Fruits. *Cooperative Extension Service,* Michigan State University.

105 Blanpied, G.D. and R.M. Smock. (1982). Storage of fresh market apples. *Cornell Univ. Info. Bull.* 191, pp.19.

106 Kingston, C.M. (1992). Maturity indices for apple and pear. *Hort. Rev.*, 13:407-432.

107 Emongor, V.E., D.P. Murr, and E.C. Lougheed. (1994). Pre-harvest factors that predispose apple to superficial scald. *Postharvest Biol. Technol.*, 4: 289-300.

108 Saure, M.C. (1996). Reassessment of the role of calcium in development of bitter pit in apple. *Austral. J. Plant Physiol.*, 23: 237-244.

109 Meheriuk, M., R.K. Prange, P.D. Lidster, and S.W. Porritt. (1994). Postharvest disorders of apples and pears. *Publ. 1737/E Agric.* Canada, Ottawa, Canada, 67 p.

110 Knee, M. (1993). Pome fruits. *In:* G.B. Seymour *et al.* (eds.), *Biochemistry of fruit ripening*. Chapman and Hall, London, pp. 325-346.

APRICOT

111 DeMartino G., Massantini R., Botondi R. and Mencarelli F. (2002) Temperature affects impact injury on apricot fruit. *Postharvest Biol. and Technol.*, 25: 145-149.

112 Bullock, R.M. and Benson, N.R. (1948). Boron deficiency in apricots. *Proc. Amer. Soc. Hort. Sci.*, 51: 199-204.

AVOCADO

113 Woolf, A.B., Watkins, C.B., Bowen, J.D., Lay-Yee, M., Maindonald, J.H., Ferguson, I.B. (1995). Reducing external chilling injury in stored 'Hass' avocados with dry heat treatments. *J. Amer. Soc. Hort. Sci.*, 120: 1050–1056.

114 Ayers, A. D., Aldrich, D. G. and Coony, J. J. (1951). Leaf Burn of Avocado: Sodium or chloride accumulation may cause burning of mature avocado leaves of Fuerte and other varieties. *Calif. Agric.*, 5(12):7.

115 Adato, I. and Gazit, S. (1974). Water-deficit stress, ethylene production, and ripening in avocado fruits. *Plant Physiol.*, 53:45-46.

116 Bower, J.P. and Cutting, J.G.M. (1988). Avocado fruit development and ripening physiology. *Hort. Rev.*, 10:229-271.

117 Schroeder, C.A. (1953). Growth and development of the 'Fuerte' avocado fruit. *Proc. Amer. Soc. Hort. Sci.*, 61:103-109.

118 Kahn, V. (1975). Polyphenol oxidase activity and browing of three avocado varieties. *J. Sci. Food Agric.*, 26: 1319-1324.

119 Van Lelyveld, L.J. and Bower, J.P. (1984). Enzyme reactions leading to avocado mesocarp discolouration. *J. Hort. Sci.*, 59: 257-263.

120 Tingwa, P.O. and Young, R.E. (1974). The effect of calcium on ripening of avocado (*Persea americana* Mill.) fruits. *J. Amer. Soc. Hort. Sci.*, 99: 540-542.

121 Eaks, I.L. (1976). Ripening, chilling injury and respiratory response of Hass and Fuerte avocado fruits at 20°C following chilling. *J. Amer. Soc. Hort. Sci.*, 101: 538-540.

BAEL

122 Saini, R.S., Singh, S. and R.P.S. Deswal (2004). Effect of micro-nutrients, plant growth regulators and soil amendments on fruit drop, cracking and yield and quality of bael (*Aegle marmelos* Corr.) under rainfed conditions. *Indian J. Hort.*, 61 (2): 175-176.

BANANA

123 Robinson, J.C. (1996). Bananas and Plantains. CABI Publishing, New York, USA.

124 Baldry, J., Coursey, D. G. and Howard, G. E. (1981). The comparative consumer acceptability of triploid and tetraploid banana fruit. *Trop. Sci.*, 23:33-66.

125 Banks, N. H. and Joseph, M. (1991). Factors affecting resistance of banana fruit to compression and impact bruising. *J. Sci. Food Agric.*, 56:315-323.

126 Lyons, J. M. (1973). Chilling injury in plants. *Annu. Rev. Plant Physiol.*, 24:445-466.

127 Meredith, D.S. (1971). Transport and storage diseases of bananas: Biology and control. *Trop. Agric.*, 48(1):35-50.

128 Morris, L.L. (1982). Chilling injury of horticultural crops: An overview. *HortScience,* 17(2):161-165.

129 New, S. and Marriott, J. (1974). Factors affecting the development of finger drop in bananas after ripening. *J. Food Technol.*, 18:241-250.

130 Wang, C.Y. (1991). Reduction of chilling injury in fruits and vegetables. *Post-harvest News and Information*, 2(3):165-168.

131 Wardlaw, C.W. (1931). Banana diseases. *Trop. Agric.*, 8(2):293-298.

132 Semple, A.J. and Thompson, A.K. (1988). Influence of the ripening environment on the development of finger drop in bananas. *J. Sci. Food Agric.*, 46:139-146.

BER

133 Abbas, M.F. (1997). Jujube. *In*: S. Mitra (ed.). *Postharvest physiology and storage of tropical and subtropical fruits.* CAB International, Wallingford, UK, pp. 405-415.

CITRUS

134. Brown G.E. and Barmore C.R. (1981) Ultrastructure of the response of citrus epicarp to mechanical injury. Botanical Gazette, 142(4), 477-481.

135. McDonald R.E., Nordby H.E. and McCollum T.G. (1993) Epicuticular wax morphology and composition are related to grapefruit chilling injury. HortScience, 28(4), 311-312.

136. Saraswathi, T., P. Rangasamy and R.S. Azhakiamanavalan. (2003). Effect of preharvest spray of growth regulators on fruit retention and quality of mandarins (*Citrus reticulata* Balanco). *South Indian. Hort.*, 51(1/6): 110-112.

137. Kanwar, J.S. and Chanana, Y.R. (1993). Granulation in citrus. In: Advances in Horticulture, Vol. 4- Fruit Crops (eds. Chadha, K.L. and Pareek, O.P.). Malhotra Publishing House, New Delhi, India, pp. 2069-2080.

138. Singh, R. and Singh, A.K. (1993). Fruit Cracking. In: Advances in Horticulture, Vol. 4- Fruit Crops (eds. Chadha, K.L. and Pareek, O.P.). Malhotra Publishing House, New Delhi, India, pp. 2119-2179.

139. Singh, R. (2000). 65 year research on citrus granulation. *Indian J. Hort.*, 58 (1-2): 112-144.

140. Addicott, F.T. (1982). Abscission. University of California Press, Berkley, USA.

141. Goren, R. (1983). Anatomical, physiological and hormonal aspects of abscission in citrus. *Hort. Reviews*, 15: 145-182.

142. Jawanda, J.S., Sinha, M.K. and Uppal, D.K. (1972). Studies on nature and periodicity of preharvest fruit drop in sweet orange. *Indian J. Hort.*, 29: 269-276.

143. Rajput, C.B.S. and Haribabu, R.S. (1999). Citriculture. Kalyani Publishers, Ludhiana.

144. Sharma, B.B. and Sharma, H.C. (1993). Fruit drop in citrus. In: Advances in Horticulture, Vol. 3- Fruit Crops (eds. Chadha, K.L. and Pareek, O.P.). Malhotra Publishing House, New Delhi, India, pp. 1185-1195.

145. Randhawa, G.S., Singh, S.P. and Mallik, R.S. (1958). Fruit cracking in some tree fruits with special reference to lemon (*Citrus limon*). *Indian J. Hort.*, 15(1) : 6-9.

146. Sharma, R.R., Saxena, R.K., Goswami, A.M. and Shukla, A.K. (2002). Effect of foliar application of calcium chloride on fruit cracking, yield and quality of Kagzi Kalan lemon. *Indian J. Hort.*, 59(2): 145-149.

147. Lavania, M.L., Mishra, K.K. and Ratnababu, G.H.V. (1986). A note on the control of fruit cracking in lemon (Citrus limon). *Maharashtra J. Hort.*, 3: 52-64.

148. Aiyappa, K.M. and Srivastava, K.C. (1967). Citrus dieback diseases in India. ICAR, New Delhi, India.

149. Dhatt, A.S. and Dhiman, J.S. (2001). Citrus decline. *Indian J. Hort.*, (Spl. Issue), 58 (1-2): 91-111.

150. Gill, K.S., Kanwar, J.S. and Singh, R. (1990). Citriculture in north western India. PAU, Ludhiana, India.

151. Gupta, S.G. and Singh, S. (2001). Citrus decline and management in NEH region. NRC Citrus, Nagpur, India.

152. Mukhopadhyay, S. (1987). Citrus decline: mapping and control project in Darjeeling district. Citrus dieback research station, Kalimpong, West Bengal, India.

153. Randhawa, G.S. and Srivastava, K.C. (1986). Citriculture in India. Hindustan Publishing Corporation, New Delhi, India.

154. Reddy, G.S. and Murti, B.D. (1990). Citrus diseases and their control. ICAR, New Delhi, India.

155. Reddy, G.S. and Murti, B.D. (1975). Citrus decline in Andhra Pradesh. APAU, Hyderabad, India.

156. Sachan, J.N. (1981). Mandarin orange decline in NEH region and its control. *Research Bulletin* No. 16.

157. Singh, S. (1997). Citrus decline: cause, remedies and future strategies. *Proc. Nat. Sump. Citriculture* held at NRC on Citrus, Nagpur, pp. 419-429.

158. Singh, S. and Naqvi, S.A.M.H. (2001). Citrus. International Book Distributing Company, Lucknow, India.

159. Awasthi, R.P. and NAuriyal, J.P. (1971). Granulation in citrus: A review. *Punjab Hort. J.*, 11(1-2): 98-104.

160. Bartholomew, E.T., Sinclair, W.B. and Ruby, E.C. (1934). Granulation of Valencia oranges. *California Citrograph,* 19: 88-89, 106-108.

161. Chanana, Y.R. (1982). Studies on the causes of citrus granulation. Ph.D. Thesis, PAU Ludhiana, India.

162. Singh, R. (1977). Some studies on citrus granulation. Ph.D. Thesis, P.G. School, IARI, New Delhi, India.

163. Subbiah, N. (1980). Studies on citrus granulation. Ph.D. Thesis, P.G. School, IARI, New Delhi, India.

164. Sharma, R.R. (2002). Citrus granulation: cause and control. *Intensive Agric.*, 42(1-2): 20-22.

165. Eckert, J.W. and Eaks, I.L. (1989). Postharvest disorders and diseases of citrus fruits. *In*: W. Reuther *et al.* (eds.), *The Citrus Industry*, Vol. V, Univ. Calif., Div. Agr. Nat. Resour. Publications, Oakland, CA, pp. 179-260.

166. Augusti, M., Martinez-Fruentes, A. and Mesejo, C. (2002). Citrus fruit quality, physiological basis and techniques of improvement. *Agrociencia,* 4(2):1-16.

167. Chen De Yan (2004). Study on the reasons causing physiological fruit drop and cracking of mandarin cv. Mashuiju. *S. China Fruits,* 33 (1): 5-6.

168. Chundawat, B.S. and Randhawa, G.S. (1972). Effect of plant growth regulators on fruit set, fruit drop and quality of Saharanpur Special variety of grapefruit. *Indian J. Hort.* 29: 277-282.

169. Garcia-Luis, A., Duarte, A.M.M., Kanduser, M. and Guardiola, J.L. (2001). The anatomy of the fruit in relation to the propensity of citrus species to split. *Scientia Hort.,* 87:33-52.

170. Holtzhausen, L.C. and Duplessis, J.A. (1970). Skin splitting of citrus fruits. *S. Africa Citrus J.,* 444: 15, 17-19.

171. Josan, J.S., Sandhu, A.S., Singh, R. and Dhillon, D.S. (1994). Effect of various soil moisture regimes on fruit cracking in lemon. *Indian J. Hort.,* 51 (2): 141-145.

172. Josan, J.S., Sandhu, A.S., Singh, R. and Dhillon, D.S. (1995). Effect of plant growth regulators and nutrients on fruit cracking in lemon. *Indian J. Hort.,* 52 (2): 121-124.

173. Monselie, S.P., M. Weiser, N. Shafir, R. Goren, E.E. Goldschmidt (1976). Creasing of orange peel. Physiology and control. *J. Hort. Sci.,* 51: 341-351.

174. Qian, K.S. and Qian, K.S. (1997). Reasons for fruit cracking in Navel oranges and its control. *S. China Fruits,* 26 (3): 12.

175. Randhawa, G.S., Singh, J.P. and Malik, R.S. (1958). Fruit cracking in some tree fruits with special reference to lemon. *Indian J. Hort.,* 15 (1): 6-9.

176. Singh, M.I. and Chohan, G.S. (1982). Effect of nutritional sprays on granulation in citrus cv. Wilking mandarin. *Punjab Hort. J.,* 22 (1-2): 59.

177. Singh, R. and Singh, R. (1981a). Effect of GA_3, planofix (NAA) and ethrel on granulation and fruit quality in Kaula mandarin. *Scientia Hort.,* 14: 315-321.

178. Singh, R. and Singh, R. (1981b). Effect of nutrient sprays on granulation and fruit quality of 'Dancy tangerine' mandarin. *Scientia Hort.,* 14 (3): 235 - 244.

DATE PALM

179. Ait-Oubahou, A. and Yahia, E.M.. (1999). Postharvest handling of dates. *Postharvest News and Information,* 10: 67N-74N.

DURIAN

180. Husin, A. and Abidin, M.Z. (1998). Durian. *In*: P.E. Shaw *et al* (eds.). *Tropical and subtropical fruits. Agscience,* Inc., Auburndale, FL, pp. 261-289.

181. Ketsa, S. (1997). Durian. *In*: S. Mitra (ed.). *Postharvest physiology and storage of tropical and subtropical fruits.* CAB International, Wallingford, UK, pp. 323-334.

182. Nanthachai, S. (1994). Durian - fruit development, postharvest physiology, handling and marketing in ASEAN. *ASEAN Food Handling Bureau*, Kuala Lumpur, Malaysia, pp. 156.

GRAPE

183. Fischer D., Craig W.L., Watada A.E., Douglas W. and Ashby B.H. (1992) Simulated in transit vibration damage to packaged fresh market grapes and strawberries. *Applied Engineering in Agriculture*, 8(3): 363-366.

184. Christensen, L. P. 1982. Waterberry - what we know today. Table Grape Seminar Proceedings, pp 12-14, Feb. 10, 1982, Dinuba. Univ. Of Calif. Cooperative Extension and California Table Grape Commission.

185. Christensen, L.P. and Boggero, J.D. (1985). A study of mineral nutrition relationships of waterberry in Thompson Seedless. *Am. J. Enol. Vitic.*, 36:57-64.

186. Chadha, K.L. and Shikhamany, S.D. (1999). The grape. Malhotra Publishing House, New Delhi, India.

187. Pandey, R.M. and Pandey, S.N. (1996). The Grape in India. ICAR, New Delhi, India.

188. Shanmugavelu, K.G. (1998). Viticulture in India. *Agro Botanica*, Bikaner, India.

189. Considine, J.A. and Kriedeman, P.E. (1972). Fruit splitting in grapes. Determination of the critical turgor pressure. *Aust. J. Agric. Res.*, 13: 17-24.

190. Zhao, S.J. and Wang, Y.F. (2002). Control of fruit cracking in Red Globe Grape. *China fruits*, 3: 53.

191. Song, G.C. and Song, N.H. (1993). Study of water status in relation to berry splitting in 'Kyoho' grape. *RDA J. Agri. Sci. Hort.*, 35 (2): 484-489.

GUAVA

164 Reyes, M.U. and Paull, R.E. (1995). Effect of storage temperature and ethylene treatment on guava (*Psidium guajava* L.) fruit ripening. *Postharvest Biol. Technol.*, 6: 357-365.

165 Ali, Z.M. and Lazan, H. (1997). Guava. *In*: S. Mitra (ed.). *Postharvest physiology and storage of tropical and subtropical fruits*. CAB International, Wallingford, UK, pp. 145-165.

166 Lazan, H. and Ali, Z.M. (1998). Guava. *In*: P.E. Shaw *et al* (eds.). *Tropical and subtropical fruits*. Agscience, Inc., Auburndale, FL, pp. 446-485.

HAZELNUTS

167 Thompson, M.M., Lagerstedt, H.B. and Mehlenbacher, S.A. (1996). Hazelnuts. *In*: J. Janick and J.N. Moore (eds) *Fruit Breeding* Vol. 3. Nuts. John Wiley and Sons, NY, pp. 125-184.

168 Richardson, D.G. (1997). The health benefits of eating hazelnuts: Implications for blood lipid profiles, coronary heart disease and cancer risks. *Acta Hort*. 445:295-300.

169 Ebraheim, K.S., Richardson, D.G. and Tetley, R.M. (1994). Effects of storage temperature, kernel intactness and roasting temperature on vitamin E, fatty acids and peroxide value of hazelnuts. *Acta Hort*., 351:677-684.

JACKFRUIT

170 Ghosh, S.P. (2000). Status report on genetic resources of jackfruit in India and SE Asia. IPGRI S Asia Office, India.

KIWI FRUIT

171 Burdon J. and Clark C. (2001). Effect of postharvest water loss on 'Hayward' kiwifruit water status. *Postharvest Biol. Technol*., 22(3): 215-225.

172 Lallu N., Rose K., Wiklund C. and Burdon J. (1999). Vibration induced physical damage in packed Hayward kiwifruit. *Acta Hort*., 498: 307-312.

173 Chauhan, J.S., Chandel, J.S. and Negi, K.S. (2001). Cultivation of kiwifruit. Department of Pomology, Dr. Y.S. Parmar University of Horticulture and Forestry, Solan (H.P.) India.

174 Arpaia, M.L., Mitchell, F.G. and Kader, A.A. (1994). Postharvest physiology and causes of deterioration. *In*: J.K. Hasey *et al*. (eds.) *Kiwifruit growing and handling*. UC DANR Publ. 3344. pp. 88-93.

175 Cheah, L.H. and Irving, D.E. (1997). Kiwifruit. *In*: S. Mitra (ed.). *Postharvest physiology and storage of tropical and subtropical fruits*. CAB International, Wallingford, UK, pp. 209-227.

176 Luh, B.S. and Wang, Z. (1984). Kiwifruit. *Adv. Food Res.*, 29: 279-305.

177 Mitchell, F.G. 1990. Postharvest physiology and technology of kiwifruit. *Acta Hort.*, 282: 291-307.

178 Sommer, N.F., J.E. Suadi, and R.J. Fortlage. 1994. Postharvest storage diseases. *In*: J.K. Hasey *et al.* (eds.). *Kiwifruit growing and handling*. UC DANR Publ. 3344, pp. 116-122.

LITCHI

179 Kadam S.S. and Deshpande S.S. (1995) Lychee. *In*: Salunke D.K. and Kadam S.S. (eds) *Handbook of Fruit Science and Technology*, Dekker, New York, 435-443.

180 Nagar P.K. (1994) Physiological and biochemical studies during fruit ripening in litchi (*Litchi chinensis* Sonn.). *Postharvest Biol. Technol.*, 4(3): 225-234.

181 Sinha, A.K., Singh, C. and Jain, B.P. (1999). Effect of plant growth substances and micronutrients on fruit set, fruit drop, fruit retention and cracking of litchi cv. Purbi. *Indian J. Hort.*, 56 (4): 309-311.

182 Verma, S.K. and Das, S.R. (1980). Effect of GA_3 spraying on the incidence of fruit cracking in litchi. *Haryana J. Hort. Sci.*, 10: 4-10.

183 Lal, H. and Mishra, S.P. (1981). Effect of irrigation frequency on fruit cracking and quality of litchi cv. Kolkata. *Punjab Hort. J.*, 21: 182-183.

184 Sharma, S.B. and Ray, P.K. (1987). Fruit cracking in litchi. A review. *Haryana J. Hort. Sci.*, 16: 11-15.

185 Opara, L.U., Studman, C.J. and Banks, N.H. (1997). Fruit skin splitting and cracking. *Hort. Rev.*, 17:216-261.

186 Teaotia, S.S. and Singh, R.D. (1970). Fruit cracking. A review. *Prog. Hort.*, 21:21-31.

187 Teaotia, S.S., Singh, R.D. and Tripathi, R.S. (1971). Studies on fruit cracking in some important sub-tropical fruits. *Prog. Hort.*, 22: 35-48.

188 Bhat, S.K., Raina, B.L., Chogtu, S.K. and Muthoo, M.K. (1997). Effect of exogenous auxin application on fruit drop and cracking in litchi. (*Litchi Chinensis* Sonn) cv. Dehradun. *Adv. Plant Sci.*, 10 (1): 83-86.

189 Bisht, G.S. and Kumar, G. (2002). Effect of different microclimatic conditions on fruit browning and cracking of litchi. *Prog. Hort.*, 34 (2): 142-145.

190 Brahmachari, V.S. and Rani, R. (2001). Effect of growth substances on productivity, cracking, ripening and quality of fruits of litchi. *The Orissa J. Hort.*, 29 (1): 44-45.

191 Chandel, S.K. and Sharma, N.K. (1992). Extent of fruit cracking in litchi and its control measure. *S. Indian Hort.*, 40 (2): 74-76.

192 Mitra, S.K. (2004). Sustainable litchi (*Litchi chinensis* Sonn.) production in West Bengal, India. *Acta Hort.*, 632: 209-212.

193 Prasad, A. and Jauhari, O.S. (1963). Effect of 2, 4, 5 – T and NAA on "drop stop" and size of litchi fruits. *Madras Agric. J.*, 50: 28-29.

194 Singh, A.K., Sharma, P., Sharma, R.M. and Tiku, A.K. (2005). Effect of plant bioregulators and micronutrients on tree productivity, fruit cracking and aril proportion of litchi (*Litchi chinensis* Sonn.) cv. Dehradun. *Haryana J. Hort. Sci.*, 34 (3-4): 220-221.

LOQUAT

195 Blumenfeld, A. (1980). Fruit growth of loquat. *J. Amer. Soc. Hort. Sci.*, 105: 747-750.

MANGO

196 Wainwright, H. and Burbage, M. B.(1989), Physiological disorders in mango (*Mangifera indica* L.) fruits: A review. *J. Hort. Sci.*, 64: 125–135.

197 Ravindra, V. and Shivashankar, S. (2004). Spongy tissue in Alphonso mango - significance of *in situ* seed germination events. *Curr. Sci.*, 87: 1045–1049.

198 Wainwright, H. and Burbage, M. B. (1989). Physiological disorders in mango (*Mangifera indica* L.) fruits: A Review. *J. Hortic. Sci.*, 64, 125–135.

199 Ravindra, V. and Shivashankar, S. (2004). Spongy tissue in Alphonso mango - significance of *in situ* seed germination events. *Curr. Sci.*, 87, 1045–1049.

200 Desai, M.C. (1966). Sponginess in mango fruits. *Junagarh Agri. College Magazine*, 5: 5-6.

201 Joshi, G.P. and Roy, S.K. (1989). Studies on spongy tissue in Alphonso mango. *Acta Hort.*, 231: 649-661.

202 Katrodia, J.S. and Bhuva, H.P. (1993). Spongy tissue in mango. In: Advances in Horticulture, Vol. 4- Fruit Crops (eds. Chadha, K.L. and Pareek, O.P.). Malhotra Publishing House, New Delhi, India, pp. 2069-2080.

203 Katrodia, J.S. and Rane, D.A. (1989). The pattern of distribution of spongy tissue in affected Alphonso fruits at different locations. *Acta Hort.*, 231: 873-877.

204 Pal, R.N. and Chadha, K.L. (1993). Black tip and internal necrosis of mango. In: Advances in Horticulture, Vol. 4- *Fruit Crops* (eds. Chadha, K.L. and Pareek, O.P.). Malhotra Publishing House, New Delhi, India, pp. 2081-2093.

205 Pal, R.N. and Chadha, K.L. (1980). Black tip disorder of mango: A review. *Punjab Hort. J.*, 20:112-121.

206 Pandey, R.M., Singh, R.N. and Sharma, Y.K. (1974). Leaf scorch in mango, a new problem. *Indian Hort.*, 19: 7-8.

207 Pandey, R.M. and Prakash, O. (1998). Mango disorders, causes and management. In: Mango Cultivation (ed. Srivastava, R.P.). International Book Distributing Company, Lucknow, India, 301-339.

208 Ram, S. (2001). Mango malformation. *Indian J. Hort.*, 58(1-2): 78-90.

209 Rane, D.A., Katrodia, J.S. and Kulkarni, D.N. (1976). Problem of spongy tissue development in mango (Mangifera indica L.) cv. Alphonso: A Review. *J. Agric. University*, 1: 89-94.

210 Singh, Z. and Dhillon, B.S. (1993). Mango malformation. In: Advances in Horticulture, Vol. 4- Fruit Crops (eds. Chadha, K.L. and Pareek, O.P.). Malhotra Publishing House, New Delhi, India, pp. 2047-2067.

211 Chadha, K.L. and Pal, R.N. (1993). Biennial bearing in mango. In: Advances in Horticulture, Vol. 4- Fruit Crops (eds. Chadha, K.L. and Pareek, O.P.). Malhotra Publishing House, New Delhi, India, pp. 2025-2046.

212 Davis, L.D. (1957). Flowering and alternate bearing. Proc. *Amer. Soc. Hort. Sci.*, 70: 545-556.

213 Rao, M.M. (1998). Problem of alternate bearing in mango: causes and management. In: Mango Cultivation (ed. Srivastava, R.P.). International Book Distributing Company, Lucknow, India, 339-362.

214 Singh, R.N. (1971). Biennial bearing in fruit trees- accent on apple and mango. Bulletin No. 30, ICAR, New Delhi, India.

215 Ram, S. (1983). Hormonal control of fruit growth and fruit drop in mango cv. Dashehari. *Acta Hort.*, 134: 169-178.

216 Randhawa, G.S. and Chadha, K.L. (1994). Fruit drop and its contrion in mango and citrus. ICAR Publication, New Delhi, India.

217 Singh, R. and Arora, K.S. (1985). Some factors affecting fruit drop in mango. *Indian J. Agric. Sci.*, 35: 196-205.

218 Singh, R.N. (1990). Mango. Indian Council of Agricultural Research, New Delhi, India.

219 Chadha, K.L. (1993). Fruit drop in mango. In: Advances in Horticulture, Vol. 3- Fruit Crops (eds. Chadha, K.L. and Pareek, O.P.). Malhotra Publishing House, New Delhi, India, pp. 1131-1165.

220 Kulkarni, V.J. (1988). Effect of post-bloom vegetative flush on fruit retention in mango. *Acta Hort.*, 231: 500-502.

221 Bajpai, P.N. and Shukla, H.S. (1978). Combating mango malformation through exogenous application of NAA. *Plant Sci.*, 10: 135-137.

222 Chadha, K.L., Pal, R.N., Parkash, O., Tandon, P.L. and Singh, H. (1979). Studies on mango malformation its causes and control. *Indian J. Hort.*, 36: 359-368.

223 Chakrabarti, D.K. and Kumar, R. (2002). Mango malformation: present status and future strategies. In: IPM systems in agriculture. Vol. 8. Aditya Books Pvt. Ltd., New Delhi, India.

224 Dang, J.K. and Daulta, B.S. (1982). Mango malformation: A Review. *Pesticides*, 16: 5-11.

225 Kumar, J., Singh, U.S. and Beniwal, S.P.S. (1993). Mango malformation. One Hundred Years of Research. *Ann. Re. Plant Physiol.*, 31: 217-232.

226 Majumdar, P.K. and Diwale, D.V. (1985). Studies on horticultural aspects of mango malformation. *Acta Hort.*, 231: 840-845.

227 Mallick, P.C. (1963). Mango malformation, symptoms, causes and cure. *Punjab Hort.* J., 3: 292-299.

228 Pandey, R.M., Rao, M.M. and Pathak, R.A. (1977). Biochemical changes associated with floral malformation in mango. *Scientia Hort.*, 6: 37-44.

229 Singh, L.B., Singh, S.M. and Nirvan, R.S. (1972). Studies on mango malformation- a review: symptoms, extent, intensity and cause. *Hort. Adv.*, 5: 197-217.

230 Ram, S. (1991). Horticultural aspects of mango malformation. *Acta Hort.*, 291: 235-252.

231 Singh, Z. and Dhillon, B.S. (1986). Effect of plant growth regulators on floral malformation, flowering, productivity and fruit quality of mango (*Mangifera indica* L.). *Acta Hort.*, 174: 315-319.

232 Srivastava, R.P. (1998). Mango malformation. In: Mango Cultivation (ed. Srivastava, R.P.). International Book Distributing Company, Lucknow, India, 363-408.

233 Tripathi, P.C. (1992). Studies on the cause and control of mango malformation. Ph.D. Thesis, G.B. Pant Uni. of Agric. and Technology, Pantnagar, India.

MANGOSTEEN

234 Sdoodee, S. and Limpun-Udom, S. (2002). Effect of excess water on the incidence of translucent flesh disorder in mangosteen (*Garcinia mangostana* L.). *Acta Hort*. 575: 813-820.

235 Osman, M. and Rahman Milan, A. (2006). Mangosteen - *Garcinia mangostana*. South Hampton Centre for Underutilised Crops, University of South Hampton, South Hampton, U.K.

OLIVE

236 Freeman, M., Uriu, K. and Hartmann, H.T. Ferguson, L., Sibbett, G.S. and Martin, G.C. (eds) (1994). The olive tree and fruit. Diagnosing and correcting nutrient problems: Olive Production Manual. University of California Publication 3353.

237 Kader, A. A., Nanos, G. D. and Kerbel, E. L. (1990). Storage potential of fresh 'Manzanillo' olives. *Calif. Agric.*, 44(3): 23-24.

PAPAYA

238 Quintana, M.E.G. and Paull, R.E. (1993) Mechanical injury during postharvest handling of 'Solo' papaya fruit. *J. Am. Soc. Hort. Sci.*, 118(5): 618-622.

239 Hine, R.B., Holtzmann, O.V. and Raabe, R.D. (1965). Diseases of papaya (*Carica papaya* L.) in Hawaii. *Hawaii. Agric. Exp. Stn. Bull.*, 136, Univ. of Hawaii, 26 pp.

PASSION FRUIT

240 Akali, S. and Maiti, C S. (2006). Status and prospects of Passion fruit industry in northeast India. Central Institute of Horticulture, Medziphema-797 106, Nagaland

PEACH

241 Cheng, G. W. and Crisosto, C. H. (1994). Development of dark discoloration on peach and nectarine (*Prunus persica* L. Batsch) fruit in response to exogenous contamination. *J. Amer. Soc. Hort. Sci.*, 119: 529-533.

242 Crisosto, C. H., Johnson, R. S., Luza, J. and Day, K. (1993). Incidence of physical damage on peach and nectarine skin discoloration development: anatomical studies. *J. Amer. Soc. Hort. Sci.*, 118: 796-800.

243 Fernandez-Trujillo, J.P., Cano, A. and Artes, F. (1998). Physiological changes in peaches related to chilling injury and ripening. *Postharvest Biol. Technol.*, 13:109–119

PEAR

244 Blanpied, D.G. (1975). Core breakdown of New York 'Bartlett' pears. *J. Amer. Soc. Hort. Sci.*, 100: 198-200.

245 Crisosto, C. H., Day, K. R., Sibbett, S, Garner, D. and Crisosto, G. (1994). Late harvest and delayed cooling induce internal browning of 'Ya Li' and 'Seuri' Chinese pears. *HortScience*, 29(6): 667-670.

246 Mitchell, F. G. and Maye, G. (1972). Watery breakdown of 'Bartlett' pear. *Calif. Agric.*, 27(5): 6-8.

247 Lallu, N. (1990). Fruit growth, handling and storage. *In*: A.G. White (ed.). Nashi, Asian pear in New Zealand, DSTR Publishing, Wellington, pp. 53-74.

248 Raese, J.T. (1989). Physiological disorders and maladies of pear fruit. *Hort. Rev.*, 11: 357-411.

PECAN NUT

249 Wood, B.W., Reilly, C.C. and Nyczepir, A.P. (2004). Mouse ear of pecan: I. Symptomatology and occurrence. *HortScience,* 39: 87-94.

250 Wood, B.W., Reilly, C.C. and Nyczepir, A.P. (2004). Mouse ear of pecan: II. Influence of nutrient applications. *HortScience,* 39: 95-100.

251 Wood, B.W., Reilly, C.C. and Nyczepir, A.P. (2004). Mouse ear of pecan: A nickel deficiency. *HortScienc,*. 39: 1238-1242.

252 Worley, R.E., and Taylor, G.G. (19720. An abnormal nut splitting problem of pecan. *HortScience,* 7: 70-71.

253 Sparks, D. (1989). Pecan nutrition-a review. *Proc. S.E. Pecan Grow. Assoc.*, 82: 101-122.

254 Sparks, D. (1976). Nitrogen scorch and the pecan. *Pecan South.*, 3: 500-501.

PERSIMMON

255 Collins, R. J. and Tisdell, J. S. (1995). The influence of storage time and temperature on chilling injury in Fuyu and Suruga persimmon (*Diospyros kaki* L.) grown in subtropical Australia. *Postharv. Biol. Technol.*, 6:149-157.

256 MacRae, E. A. (1987). Development of chilling injury in New Zcaland grown Fuyu persimmon during storage. *N.Z.J. Exp. Agric.*, 15: 333-344.

PINEAPPLE

257 Paull, R.E., Reyes, M.E.Q. (1996). Preharvest weather conditions and pineapple fruit translucency. *Scientia Hort.*, 66: 59–67.

POMEGRANATE

258 Ben-Arie, R. and Or, E. (1986). The development and control or husk scald on 'Wonderful' pomegranate fruit during storage. *J. Hort. Sci.,* 111: 395-399.

259 Elyatem, S. M. and Kader, A. A. (1984). Post-harvest physiology and storage behavior of pomegranate fruits. *Scientia Hort.,* 24: 287-298.

260 Kader, A. A., Chordas, A. and Elyatem, S. (1984). Responses of pomegranates to ethylene treatment and storage temperature. *Calif. Agric.,* 38(7-8): 14-15.

261 Prasad, R.N., Banker, G.J. and Vashishtha, B.B. (2003). Problems and prospects of pomegranate cultivation in arid region. *Scientific Hort.,* 8: 25-30.

262 Singh, D.B., Kingsly, A.R.P. and Jain, R.K. (2006). Controlling fruit cracking in pomegranate. *Indian Hort.,* 51(1): 14, 32.

263 Singh, R.P., Sharma, Y.P. and Awasthi, R.P. (1990). Influence of different cultural practices on pre-mature fruit cracking of pomegranate. *Prog. Hort.,* 22 (1-4): 92-96.

264 Sonawane, P.C., Desai, U.T., Chaudhari, S.M. and Deore, B.P. (1994). Extent of fruit loss in pomegranate due to biotic and abiotic factors. *J. Maharashtra Agric. Univ.,* 19 (1): 161-162.

STRAWBERRY

265 Sharma, R.R. (2002). Growing Strawberries. International Book Distributing Co., Lucknow, India.

266 Hancock, J.F. (1999). Strawberries. CABI Publishing, New York, USA.

267 Sharma, V.P. and Sharma, R.R. (2004). The Strawberry. ICAR, New Delhi, India.

268 Khokhar, U.U., Prashad, J. and Sharma, M.K. (2004). Influence of growth regulators on growth, yield and quality of strawberry cv. Chandler. *Haryana J. Hort. Sci.,* 33 (3 & 4): 186-188.

WALNUT

269 Maté, J.I., Saltveit, M.E. and Krochta, J.M. (1996). Peanut and walnut rancidity: Effects of oxygen concentration and RH. *J. Food Sci.*, 61: 465-468, 472.

270 Tomar, C.S. and Singh, N. (2007). Effect of application of nutrients and bio-regulators on growth, fruit set, yield and nut quality of walnut. *Indian J. Hort.*, 64(3): 271-273.

Zeitfracht Medien GmbH
Ferdinand-Jühlke-Straße 7
99095 Erfurt, Deutschland
produktsicherheit@kolibri360.de